ISSN 0323-3898

Volume 10 · 1979 · Number 1

MATHEMATISCHE OPERATIONSFORSCHUNG UND STATISTIK

optimization

OPERATIONS RESEARCH AND MATHEMATICAL STATISTICS

ИССЛЕДОВАНИЕ ОПЕРАЦИИ И МАТЕМАТИЧЕСКАЯ СТАТИСТИКА

RECHERCHE OPERATIONELLE ET STATISTIQUE MATHEMATIQUE

1

Akademie-Verlag · Berlin

:VP 20,– M 32 585

Instructions for Authors

Submission of paper. Papers to be submitted for publication or books to be reviewed should be sent to

PROFESSOR K.-H. ELSTER
Chief Editor, Optimization
TH Ilmenau
Sektion Mathematik, Rechentechnik und ökonomische Kybernetik
DDR – 63 Ilmenau, Ehrenberg, Block G

Papers may be written in English, French, German or Russian. It is assumed, that the papers have not been published previously. The original (or xerox copy) should be submitted with two additional copies, which will not be returned to the author. Authors are held fully responsible for all information included in their papers.

Preparation of manuscripts. Manuscripts should be typewritten, as least double-spaced with wide margins at all four sides. Before having a paper typed for submission, please examine a recent issue of Optimization carefully and follow the arrangement of formulae, references, tables etc. Heading should be numbered by arabic numbers. The exact position of mathematical characters must be absolutely unambiguous. Distinguish between "oh" and "zero", "el" and "one", "kappa" and "kei", etc.

Text (with exception of mathematical symbols and formulae) may be stressed manually in the following ways

Bold	straight black underline
Italics	waved blue underline
Spaced type	dotted underline
Small capitals	waved red underline

The following script annotation rules shall be observed for mathematical symbols and formulae:

Greek letters (excluding summation and product signs) should

be underlined		with a red line
bold roman types	$(\mathbf{A}, \mathbf{B}, \ldots)$	with two black lines
bold italics	$(\boldsymbol{A}, \boldsymbol{B}, \ldots)$	with two blue lines
gothic types	$(\mathfrak{A}, \mathfrak{B}, \ldots)$	with a green line
bold gothic types	$(\boldsymbol{\mathfrak{A}}, \boldsymbol{\mathfrak{B}}, \ldots)$	with two green lines
straight type	$(\mathsf{A}, \mathsf{B}, \ldots)$	with a violet line
script	$(\mathcal{A}, \mathcal{B}, \ldots)$	with a brown line

The author may list on a separate sheet entitled "Hits for the printer" information which will help to avoid misinterpretations.

Summary and key words. A selfcontained summary, if possible in three of the languages mentioned, should be typed on a separate page. Its length should not exceed 150 words. It should be followed by a list of not more than 10 key words indicating the contents of the paper.

Tables and illustrations. Each table and each illustration should be prepared on a separate sheet. They should be numbered consecutively throughout the paper and be refered to in the text as Table 1, Fig. 1 etc. Their approximate position for insertion should be indicated in the margin of the manuscript. The number of illustrations should be kept as low as possible.

Footnotes. Should be reduced to minimum and, where possible, should be replaced by remarks in the text. The first footnote should be the present addresses of the authors on the first page.

Proofs and reprints. Authors will receive page proofs (in two sets). One corrected set of proofs should be returned to the Editor within on week. Authors will be charged for alterations in excess 10% of the cost of composition.

75 reprints are offered free; up two 250 additional reprints are provided at cost price.

optimization

MATHEMATISCHE OPERATIONSFORSCHUNG UND STATISTIK

SERIES OPTIMIZATION

Volume 9
Number 1–4
Page 1–640

January/December
1978

Published at
Sektion Mathematik, Rechentechnik
und ökonomische Kybernetik ·
Technische Hochschule Ilmenau

AKADEMIE-VERLAG · BERLIN

Zeitschrift „Mathematische Operationsforschung und Statistik – Series Optimization"

Herausgeber: Akademie-Verlag, DDR – 108 Berlin, Leipziger Straße 3–4; Fernruf 2236221 oder 2236229; Telex-Nr. 114420; Bank: Staatsbank der DDR, Berlin, Kto-Nr.: 6836-26-27012

Chefredakteur: Prof. Dr. Karl-Heinz Elster
Verantwortlicher Redakteur: Dr. Rüdiger Reinhardt

Anschrift der Redaktion: Sektion Mathematik, Rechentechnik und ökonomische Kybernetik der TH Ilmenau, DDR – 63 Ilmenau, Postfach 327

Veröffentlicht unter der Lizenznummer 1586 des Presseamtes beim Vorsitzenden des Ministerrates der Deutschen Demokratischen Republik

Gesamtherstellung: VEB Druckerei „Gottfried Wilhelm Leibniz",
DDR – 445 Gräfenhainichen

Erscheinungsweise: Die Zeitschrift „Mathematische Operationsforschung und Statistik – Series Optimization" erscheint jährlich in einem Band mit 4 Heften. Bezugspreis je Band 120,– M zuzüglich Versandspesen (Preis für die DDR 80,– M); Preis je Heft 30,– M (Preis für die DDR 20,– M)

Bestellnummer 1093

AN (EDV) 40234

Contens of Volume 9 (1978)

Original papers

Book Reviews

Informations

Announcement

Authors[1]

[1] Authors of original papers (number of page is printed bold), authors of reports, authors and editors of books (+), and referees of books (+ +)

ISSN 0323-3898

optimization

MATHEMATISCHE OPERATIONSFORSCHUNG UND STATISTIK

SERIES OPTIMIZATION

Volume 10
1979
Number 1

Published at
Sektion Mathematik, Rechentechnik
und ökonomische Kybernetik
Technische Hochschule Ilmenau

AKADEMIE-VERLAG · BERLIN

Editorial Policy

The journal publishes theoretical and applied papers related to the fields of linear, nonlinear, stochastic, parametric, discrete and dynamic programming, control theory, game theory, mathematical modelling of economical and technological processes, inventory, queueing theory, reliability, renewal theory, network methods etc. It is expected, that the papers give interesting and novel contributions to the theory at a high mathematical level or a stimulating presentation of ideas in foundations or applications. All papers should begin with a clear introduction including practical or theoretical consequences and connections to previous works. The trend to highest generality should not obscure the essential ideas.
A special section will be devoted to survey papers on theory and methods in interesting areas of mathematical programming and optimization techniques. Furthermore, the journal publishes proceedings of conferences, book reviews and announcements on related topics. Experts will be asked to give an opinion on each paper submitted for publication.

Math. Operationsforsch. Statist., Ser. Optimization, Vol. 10 (1979) No. 1, 3–16

Relatively Convex Sets and Functions in R^n

JOSEF NEDOMA[1]

Summary: In order to extend the concept of a convex set to the subsets of nonlinear manifolds, the so-called relative convexity to a convex cone is defined. The relative convexity provides the possibility to generalize the concept of a convex function in such a manner that it can be used to formulate the criteria of global optimality for a class of nonconvex optimization problems.

In order to extend the concept of a convex set to the subsets of nonlinear manifolds, the so-called relative convexity to a convex cone is defined in [4]. There is a relation to the similar concept of cone convexity introduced by YU in [7], the purpose and the properties of these two concepts, however, are different. The relative convexity provides the possibility to generalize the concept of a convex function in such a manner that it can be used to formulate the criteria of global optimality for a class of nonconvex optimization problems.

I. Regular Sets

We shall introduce the *concept of regularity* with respect to a given (linear) subspace of R^n.

The closure, the linear hull, the affine hull and the convex hull of a set X are denoted by $\mathrm{cl}\,X$, $\mathrm{lin}\,X$, $\mathrm{aff}\,X$ and $[X]$ respectively. Further, $\mathrm{ray}\,X=\{z=tx \mid x\in X, t>0\}$. As it is usual, we refer to a set K as to a cone if $K=\{0\}\cup \mathrm{ray}\,K$.

We introduce the so called *difference cone* of a set X as follows:

$$\mathrm{dico}\,X=\mathrm{cl}\,\mathrm{ray}\,(X-X)\,.$$

The difference cone, of course, is not convex in general. Let us denote ${}^0X=X\setminus\{0\}$.

Definition (1.1): Let L be a subspace. A set M is called *L-regular* if

$$ {}^0L\cap \mathrm{dico}\,M=\emptyset\,. \tag{1.2}$$

Lemma (1.3): *Let M be L-regular. Then the intersection $(x+L)\cap M$ contains exactly one point for every $x\in M+L$ and it is empty in the other case.*

[1] Ekonomicko-matematiká laboratoř při Ekonomickém ústavu ČSAV. Politických vězňů 7, ČS-11173 Praha 1.

Proof: The intersection $(x+L)\cap M$ is non-empty if and only if there is a $y\in M$ such that $x\in y+L$, i.e. $x\in M+L$. For arbitrary $y^1, y^2\in M\cap(x+L)$ we have $y^2-y^1\in L\cap(M-M)$ which implies $y^1=y^2$. ▮

Lemma (1.4): *Let L be a subspace and K be a closed cone such that ${}^0L\cap K=\emptyset$. Then there is a number $\eta>0$ such that for every $w\in{}^0L$*

$$C_\eta(w)\cap K=\emptyset\,, \tag{1.5}$$

where $C_\eta(w)$ is defined as follows:

$$C_\eta(w)=\{u \mid \langle u, w\rangle>(1-\eta)\ |u|\ |w|\}\,.$$

Proof: Consider the unit sphere $S=\{u \mid |u|=1\}$ and put

$$S_1=L\cap S \quad \text{and} \quad S_2=K\cap S\,.$$

S_1, S_2 are disjoint compact sets. Consequently

$$d=\inf\{\varrho(u, v) \mid u\in S_1, v\in S_2\}>0\,,$$

where ϱ is the Euclidean metric.
For any $u, v\in S$

$$\varrho^2(u, v)=\langle u-v, u-v\rangle=2\ (1-\langle u, v\rangle)$$

and therefore for every $u\in S_1$, $v\in S_2$

$$2\ (1-\langle u, v\rangle)\geqq d^2\,,$$

or

$$\langle u, v\rangle\leqq 1-\eta, \quad \text{where} \quad \eta=\frac{d^2}{2}>0\,. \tag{1.6}$$

Choose arbitrarily $w\in{}^0L$, $z\in K$. Then there exist $u\in S_1$, $v\in S_2$ such that $w=t_1u$ and $z=t_2v$ with $t_1>0$, $t_2\geqq 0$. Since $t_1=|w|$ and $t_2=|z|$, the relation (1.6) yields

$$\langle w, z\rangle=t_1t_2\langle u, v\rangle\leqq(1-\eta)\ |w|\ |z|\,,$$

which implies (1.5). ▮

Now, let M be an L-regular set and consider the mapping

$$\psi(x)=(x+L)\cap M, \quad x\in M+L\,. \tag{1.7}$$

The defining relation (1.7) is justified by lemma (1.3).

Let us recall that an arbitrary mapping $\varphi\colon A\to R^n$ $(A\subset R^n)$ is called to be satisfying the Lipschitz condition if there exists a constant β such that

$$|\varphi(z)-\varphi(z^0)|\leqq\beta\ |\ z-z^0| \quad \forall z, z^0\in A\,. \tag{1.8}$$

Theorem (1.9): *Let L be a subspace and M be an L-regular set. Then ψ defined by (1.7) is a Lipschitz-type mapping.*

Proof: 1. Let us denote $M'=(M+L)\cap L^\perp$, where $L^\perp$ is the orthogonal complement of L and choose arbitrarily $x^0, x\in M'$. Further let $y^0=\psi(x^0)$, $y=\psi(x)$ and

$$u=x-x^0, \quad v=y-y^0, \quad w=v-u=(y-x)-(y^0-x^0)\,. \tag{1.10}$$

Clearly $w \in L$. Suppose meanwhile that $w \neq 0$. According to lemma (1.4) there exists $\eta > 0$ such that

$$C_\eta(w) \cap \operatorname{dico} M = \emptyset .$$

Hence the point $y = \psi(x)$ cannot lie in $(y^0 + C_\eta)$. Consequently $v = y - y^0 \notin C_\eta$ so that

$$\langle w, v \rangle \leqq (1-\eta)\, |w|\, |v| . \tag{1.11}$$

But (1.10) implies $u \perp w$ and

$$v = u + w \tag{1.12}$$

so that $\langle w, v \rangle = |w|^2$. Putting (1.12) into (1.11) we obtain

$$|w|^2 \leqq (1-\eta)\, |w|\, (|u| + |w|)$$

and hence

$$|w| \leqq \frac{1-\eta}{\eta}\, |u| . \tag{1.13}$$

The identity (1.12) yields $|v| \leqq |u| + |w|$. Thus we can modify (1.13) as follows:

$$|y - y^0| = |v| \leqq \frac{1}{\eta}\, |x - x^0| . \tag{1.14}$$

This relation has been derived provided that $w \neq 0$. If $w = 0$, however, (1.14) is satisfied trivially for every η lying in the interval $0 < \eta \leqq 1$.

The number η is independent of w. Therefore, it does not depend on the choice of the points x^0, $x \in M'$.

2. Consider arbitrary points $z, z^0 \in M + L$. If we denote

$$x^0 = (z^0 + L) \cap L^\perp \quad \text{and} \quad x = (z + L) \cap L^\perp , \quad \text{then} \quad |x - x^0| \leqq |z - z^0| .$$

Since $\psi(x^0) = \psi(z^0)$, $\psi(x) = \psi(z)$, (1.14) yields:

$$|\psi(z) - \psi(z^0)| \leqq \frac{1}{\eta}\, |z - z^0| .$$

Thus ψ satisfies the LIPSCHITZ condition with the constant $\beta = \frac{1}{\eta} > 0$. ∎

Corollary (1.15): *Under hypotheses of the theorem* (1.9) ψ *is uniformly continuous.*

Consider now the parallel projection $\pi \colon M \to L^\perp$. π can also be spoken about as the parallel projection with the direction L. For every $x \in M$

$$\pi(x) = (x + L) \cap L^\perp \tag{1.16}$$

so that $M' = \pi(M) = (M + L) \cap L^\perp$. Clearly $M + L = M' + L$.

Due to lemma (1.3) there exists the inverse mapping π^{-1}. It is evident that π^{-1} can be expressed by restricting the mapping ψ defined by (1.7) to the set M'. Being a projection π is continuous and π^{-1} is also continuous by corollary (1.15). This yields:

Theorem (1.17): *Let L be a linear subspace and M be an L-regular set. Then the orthogonal projection* π *defined by* (1.16) *stands for a homeomorphism of M onto* $M'=\pi(M)$.

The theorem (1.9) holds also in the converse direction:

Theorem (1.18): *If* (1.7) *is a* LIPSCHITZ*-type mapping, then M is L-regular.*

Proof: The condition $|\psi(x)-\psi(x^0)|\leqq\beta\,|x-x^0|$ means that every vector $v\in M-$ $-M$ satisfies

$$|v|\leqq\beta\,|\pi(v)|\,. \tag{1.19}$$

The set K of all v's satisfying (1.19) is a closed cone. $\pi(L)=0$ yields $K\cap L=\{0\}$ which proves the L-regularity of M because dico $M\subset K$. ∎

II. Relatively Convex Sets

For simplicity, we shall call a set C-regular if it is L-regular for $L=\text{lin}\ C$.

Definition (2.1): Let C be a closed convex cone. A set M is called *relatively convex with respect to C* (briefly: *C-convex*) if
1^0. M is C-regular;
2^0. $M+C$ is a convex set.

Definition (2.2): If M is a C-convex set then C is called a *directing cone of M.*

Let us notice that a set is convex if and only if it is $\{0\}$-convex.

Let us denote $X^{[2]}=\{[a,b]\mid a,b\in X\}$.

The following theorem concerns the problem of verifying the second condition of the definition (2.1).

Theorem (2.3): *Let C be a convex cone. Then* $M+C$ *is convex if and only if*

$$M^{[2]}\subset M+C\,. \tag{2.4}$$

Proof: Since $M^{[2]}\subset[M]\subset[M+C]$, (2.4) is a necessary condition for the convexity of $M+C$. We shall prove the "if" part of the theorem. Suppose that (2.4) holds and choose arbitrarily $x^1, x^2\in M+C$ and $t\in[0,1]$. There exist points $y^i\in M\cap(x^i-C)$, $i=1,2$. Evidently $y\in(1-t)\,y^1+ty^2\in M^{[2]}\cap(x-C)$, where $x=$ $=(1-t)\,x^1+tx^2$. Hence $x\in y+C\subset M^{[2]}+C\subset M+C+C=M+C$, which means that $M+C$ is convex. ∎

Corollary (2.5): $M+C$ *is convex if and only if* $[M]\subset M+C$.

A subspace L is a closed convex cone by definition. Thus we can speak about L-convex sets. Apparently, a C-convex set M is $\tilde{C}$-convex for every closed convex cone $\tilde{C}$ satisfying $C\subset\tilde{C}\subset\text{lin}\ C$. It follows immediately from the identity $M+\tilde{C}=$ $=(M+C)+\tilde{C}$.

Theorem (2.6): *Every relatively convex set is homeomorphic to a convex set.*

Proof: Consider a C-convex set M and let M' be the set defined as $M'=(M+L)\cap L^{\perp}$, where $L=\text{lin } C$. M is L-convex and therefore $M+L$ is convex. Hence M' is also convex. Since M is L-regular, M' is homeomorphic to M by the theorem (1.17). ▮

Corollary (2.7): *Every relatively convex set is connected.*

Let M be L-convex and $M'=(M+L)\cap L^{\perp}$. It is natural to define the dimension, the relative interior and the relative boundary of M as follows:

$$\dim M=\dim M', \quad \text{ri } M=\psi(\text{ri } M') \quad \text{and} \quad \text{rel bd } M=\text{cl } M\backslash \text{ri } M .$$

Theorem (2.8): *Let M be a C-convex set. Then*

$$\dim M=\dim (M+C)-\dim C .$$

Proof: For $L=\text{lin } C$ and $M'=\pi(M)$, we have

$$\begin{aligned}\dim (M+C)&=\dim (M+L)=\dim (M'+L)=\dim M'+\dim L\\ &=\dim M+\dim C .\end{aligned}$$ ▮

Theorem (2.9): *The relative interior of a C-convex set is C-convex.*

Proof: Let M be C-convex and $L=\text{lin } C$. For every $x\in[\text{ri } M]$ there exists a $y\in M$ such that $x\in y+C$. But $y\in([\text{ri } M]+L)\cap M=\psi(\text{ri } M')=\text{ri } M$. It means that $[\text{ri } M]\subset \text{ri } M+C$ so that $\text{ri } M+C$ is convex according to corollary (2.5). The C-regularity of ri M is evident. ▮

Let us denote $\text{dir } M=\text{lin } (M-M)$.

Theorem (2.10): *Consider a set M and convex cones C, $\tilde{C}$. Let*

1⁰. *M be C-convex;*

2⁰. $(\text{lin } \tilde{C})\cap \text{dir } (M+C)=\{0\}$.

Then $\tilde{M}=M+\tilde{C}$ is C-convex as well.

Proof: We have $\tilde{M}-\tilde{M}\subset M-M+\text{lin } \tilde{C}$. Therefore the intersection of $L=\text{lin } C$ with dico $\tilde{M}$ is a subset of $L\cap(\text{dico } M+\text{lin } \tilde{C})=X$. The inclusion $L\subset \text{dir } (M+C)$ yields $X\subset \text{dir } (M+C)$ and hence $X=L\cap \text{dico } \tilde{M}=\{0\}$. It means that $\tilde{M}$ is C-regular. In addition, $\tilde{M}+C=(M+C)+\tilde{C}$ is a convex set. ▮

We shall assume R^n to be a subspace of R^{n+1}, i.e. $R^n\equiv R^n\times\{0\}$. Having a convex cone $C\subset R^n$ we can thus speak about C-convex sets in R^{n+1}.

Corollary (2.11): *Let $C\subset R^n$ be a closed convex cone and $M\subset R^n$ be a C-convex set. Then the set*

$$M'=\{(x, \mu) \mid x\in M, \mu\in R\}$$

is C-convex in R^{n+1}.

Proof: Choose $\tilde{C}=\text{lin } e^{n+1}$ (where $e^{n+1}=(0_n, 1)$) and apply the theorem (2.10). ▮

The intersection of C-convex sets is not C-convex in general. Nevertheless, we can formulate an analogous statement. First we give a lemma.

Lemma (2.12): *Let A, M_1, $M_2\subset R^n$ be given such that the intersection*

$(x-A)\cap(M_1\cup M_2)$ *contains at most one point for any* $x\in R^n$. *Then*

$$M_1\cap M_2+A=(M_1+A)\cap(M_2+A)\,. \tag{2.13}$$

Proof: If some of the sets M_1, M_2, A is empty, then (2.13) stands for the trivial identity $\emptyset=\emptyset$. Assume that all three sets are non-empty. Put $D=M_1\cap M_2+A$.
1. Apparently $D\subset(M_i+A)$ $\forall i=1,2$ and therefore

$$D\subset(M_1+A)\cap(M_2+A)\,. \tag{2.14}$$

2. Choose $x\in(M_1+A)\cap(M_2+A)$. Then there exist $y^1\in M_1$, $y^2\in M_2$, $u^1,u^2\in A$ such that $x=y^1+u^1=y^2+u^2$. Hence $y^1,y^2\in(x-A)\cap(M_1\cup M_2)$ and therefore $y^1=y^2$ according to the hypothesis of the lemma. Thus $y^1,y^2\in M_1\cap M_2$ so that $x\in D$. Thereby the inclusion $D\supset(M_1+A)\cap(M_2+A)$ is verified which together with (2.14) proves (2.13). ∎

Theorem (2.15): *Let* M_1, M_2, P *be C-convex sets such that* $M_1, M_2\subset P$. *Then the intersection* $M=M_1\cap M_2$ *is C-convex as well.*

Proof: M is L-regular since $M\subset P$. According to lemma (1.3), the intersection $(x-C)\cap(M_1\cup M_2)$ contains at most one point for every $x\in R^n$. Using lemma (2.12) we can thus conclude that the set $M_1\cap M_2+C=(M_1+C)\cap(M_2+C)$ is convex being the intersection of two convex sets. ∎

As we have already mentioned, a given relatively convex set can have more than one directing cone in general. This topic will be particularly discussed in the following two theorems:

Theorem (2.16): *Let M be a C-convex set and C′ be a closed convex cone such that* $(\operatorname{lin} C')\cap\operatorname{dir}(M+C)=\{0\}$. *Then M is* $\tilde C$*-convex for* $\tilde C=C+C'$.

Proof: Consider $\tilde L=\operatorname{lin}\tilde C$ and $X=\tilde L\cap\operatorname{dico} M$. Evidently $X\subset\hat R\cap\operatorname{dir}(M+C)=$ $=L=\operatorname{lin} C$ and therefore $X=\tilde L\cap\operatorname{dico} M=\{0\}$ which proves the $\tilde C$-regularity of M. It remains to notice that $M+\tilde C=(M+C)+C'$ is a convex set. ∎

Corollary (2.17): *Every k-dimensional relatively convex set possesses an* $(n-k)$*-dimensional directing cone.*

Proof: Let M be a k-dimensional C-convex set. By virtue of theorem (2.16), M is relatively convex with respect to $\tilde C=C+L'$, where L' is the orthogonal complement to $L=\operatorname{dir}(M+C)$. Clearly $\operatorname{dir}(M+\tilde C)=R^n$ and hence $\dim\tilde C=$ $=\dim(M+\tilde C)-\dim M=n-k$. ∎

Lemma (2.18): *The identity*

$$(X+Y)\cap\operatorname{aff} X=X+(Y\cap\operatorname{dir} X)$$

holds for arbitrary $X, Y\subset R^n$.

Proof: *Let us denote* $D=(X+Y)\cap\operatorname{aff} X$
1. For an arbitrary $u\in Y\cap\operatorname{dir} X$, $X+u\subset D$ and hence

$$X+(Y\cap\operatorname{dir} X)\subset D\,. \tag{2.19}$$

2. Choose $z \in D$. a) Since $D \subset X+Y$, there exist $x \in X$, $y \in Y$ such that $z=x+y$.
b) $D \subset \operatorname{aff} X = x + \operatorname{dir} X$. Thus there exists a $v \in \operatorname{dir} X$ such that $z=x+v$.
Comparing a), b), we obtain $y=v=z-x$. Hence $y \in Y \cap \operatorname{dir} X$, i.e. $z \in X+(Y \cap \cap \operatorname{dir} X)$. It yields

$$D \subset X + (Y \cap \operatorname{dir} X)$$

which completes the proof. ▮

Theorem (2.20): *Let M be a C-convex set and let $\tilde{C} = C \cap \operatorname{dir} M$. Then M is $\tilde{C}$-convex.*

Proof: The $\tilde{C}$-regularity of M is evident. Applying lemma (2.18) we obtain $M+\tilde{C} = (M+C) \cap \operatorname{aff} M$ which is a convex set. ▮

Corollary (2.21): *Every relatively convex set M possesses a directing cone $C \subset \subset \operatorname{dir} M$.*

Theorem (2.22): *A k-dimensional relatively convex subset of a k-dimensional affine manifold is convex.*

Proof: Let W^k be a k-dimensional affine manifold and let $M \subset W^k$ be relatively convex, $\dim M = k$. According to corollary (2.21), M possesses a directing cone $C \subset \operatorname{dir} W^k$. Then $\dim C \leqq \dim W^k - \dim M = 0$, i.e. $C = \{0\}$. ▮

The following two theorems give sufficient conditions ensuring that a directing cone of a relatively convex set M is kept for a subset of M.

Theorem (2.23): *Let C be a directing cone of M and $\tilde{M}$ be a subset of M such that $\tilde{M}$ is L-convex for $L = \operatorname{lin} C$. Then C is a directing cone of $\tilde{M}$.*

Proof: The C-regularity of $\tilde{M}$ is trivial. Choose arbitrarily $x \in [\tilde{M}]$. Since $\tilde{M}$ is L-regular, the intersection $(x+L) \cap \tilde{M}$ contains exactly one point; let us denote it by z. We have $x \in [\tilde{M}] \subset [M] \subset M+C$ and hence $z \in (x-C) \cap M = (x-C) \cap \tilde{M}$. It means that $[\tilde{M}] \subset \tilde{M}+C$ so that $\tilde{M}+C$ is convex. ▮

Definition (2.24): Let L be a subspace. A convex set V will be called an *L-cylinder* if $V+L=V$.

Lemma (2.25): *Let V be an L-cylinder. Then*

$$V \cap A + L = V \cap (A+L) \tag{2.26}$$

holds for any $A \subset R^n$.

Proof: Let us put $D = V \cap (A+L)$.
1. Evidently $V \cap A + L \subset V + L = V$ and further $V \cap A + L \subset A + L$ so that

$$V \cap A + L \subset D\,. \tag{2.27}$$

2. Choose $x \in D$. There exist $y \in A$, $u \in L$ such that $x=y+u$. Since $x \in V$, $y = x-u \in \in V-L = V+L = V$. Hence $y \in V \cap A$. It means that $x \in V \cap A + L$ and therefore $V \cap A + L \supset D$. This inclusion together with (2.27) yields (2.26). ▮

Theorem (2.28): *Let M be a C-convex set and V be an L-cylinder, where $L = \text{lin } C$. Then the intersection $\tilde{M} = M \cap V$ is a k-dimensional C-convex set, where*

$$k = m - r, \quad m = \dim L, \quad r = \dim (V \cap (M + L)) .$$

Proof: 1. At first, we shall show that $\tilde{M}$ is L-convex. To prove this, it is only to show that the sum $\tilde{M} + L$ is convex. Putting $A = M$, we have by lemma (2.25) $\tilde{M} + L = V \cap (M + L)$. Thus $\tilde{M} + L$ is convex being the intersection of two convex sets.
2. $\dim \tilde{M} = \dim (\tilde{M} + L) - \dim L = k$.
3. Applying the preceding theorem, we obtain that $\tilde{M}$ is C-convex. ∎

Theorem (2.29): *The closure of a C-convex set is C-convex.*

Proof: Let M be C-convex. 1. Since ray cl $M \subset$ cl ray M, the C-regularity of M yields the C-regularity of cl M.
2. Choose $x^1, x^2 \in \text{cl } M$ and let X be an arbitrary bounded convex set containing x^1, x^2. Then $\tilde{M} = M \cap (X + \text{lin } C)$ is a bounded C-convex set according to theorems (1.9), (2.28). The sum cl $\tilde{M} + C$ is closed because cl $\tilde{M}$ is compact and C is closed. Hence cl $\tilde{M} + C = \text{cl } (\tilde{M} + C)$ is a convex set. Consequently $[x, y] \subset \text{cl } \tilde{M} + C \subset \text{cl } M + C$ and therefore cl $M + C$ is convex by theorem (2.3).

III. C-Arcs

Definition (3.1): A closed one-dimensional C-convex set is called a (non-degenerated) *C-arc*.

Note (3.2): Every one-point set will be considered as a degenerated C-arc.

By virtue of theorem (2.6), every non-degenerated C-arc ω can be expressed as a homeomorphic image of the interval $[0, 1]$. Consider thus a homeomorphism $\varphi: [0, 1] \to \omega$. The points $x^1 = \varphi(0)$, $x^2 = \varphi(1)$ are called extreme points of ω. If we admit also degenerated arcs, then the mentioned mapping φ is not generally homeomorphic; it is only continuous. We also say that the arc ω connects the points x^1, x^2.

Theorem (3.3): *Let M be a C-convex set. Then for every points $x^1, x^2 \in M$ there exists a unique C-arc connecting x^1, x^2 in M.*

Proof: 1. First we shall prove the existence.
a) If $x^1 = x^2$ then there is nothing to prove.
b) Let $x^1 \neq x^2$. Denote

$$L = \text{lin } C, \quad V = ([x^1, x^2] + L), \quad \omega = M \cap V .$$

The set V is an L-cylinder and $\dim V = 1 + \dim L$ since $x^2 - x^1 \notin L$. According to theorem (2.28) ω is a C-arc. It can be given by a parametric expression:

$$\omega(t) = \psi\left((1 - t)\, x^1 + t x^2\right) \tag{3.4}$$

where ψ is defined by (1.7). Clearly $\omega(0)=x^1$, $\omega(1)=x^2$. (3.4) is a homeomorphism of [0, 1] onto M. Thus ω connects x^1, x^2.

2. Let $\bar{\omega}\subset M$ be an arbitrary C-arc connecting x^1, x^2. Then $\bar{V}=\bar{\omega}+L$ is an L-cylinder and $\bar{\omega}=M\cap\bar{V}$. Since $x^1, x^2\in\bar{\omega}$ we have $[x^1, x^2]\subset\bar{V}$ and hence $V\subset\bar{V}$. Therefore $\omega\subset\bar{\omega}$ which implies $\bar{\omega}=\omega$ since ω, $\bar{\omega}$ have the same extreme points. The uniqueness is proved. ▮

Theorem (3.5): *Let C be a closed convex cone and let a set M satisfy the following conditions:*

1⁰. *M is C-regular;*

2⁰. *For every $x^1, x^2\in M$ there exists a C-*arc *$\omega_{MC}(x^1, x^2)\subset M$. Then M is C-convex.*

Proof: The inclusion $[x^1, x^2]\subset(\omega_{MC}(x^1, x^2)+C)\subset M+C$ holds for every x^1, $x^2\in M$. It means that $M^{[2]}\subset M+C$. Thus, it suffices to apply theorem (2.3). ▮

IV. Semiconvex Sets

The condition of C-regularity may seem to be too strict. Indeed, it can be weakened a little provided that an additional assumption is imposed on the cone C.

Let us recall that a convex cone K is called pointed if $K\cap(-K)=\{0\}$. It can be easily shown that a closed convex cone C is *pointed* if and only if there exist linearly independent vectors $v^1, \ldots, v^m\in \text{lin } C$ such that

$$C\subset\{x=t_1v^1+\ldots+t_mv^m \mid t_i\geqq 0, \quad i=1, \ldots, m\}\,. \tag{4.1}$$

Definition (4.2): Let C be a pointed closed convex cone. A set M is called *C-semiconvex* if

1⁰. $(M-M)\cap\text{lin } C=\{0\}$;

2⁰. $M+C$ is a convex set.

Let us note, that the closure of a C-semiconvex set is not C-semiconvex in general. For example, consider the following set in R^2:

$$M=\text{ri }[(0, 0), (0, 1)]\cup\{(1, 1)\}\,.$$

Evidently, M is C-semiconvex for the one-dimensional cone C which is generated by the vector $u=(1, 0)$. The closure of M, however, can not be C-semiconvex because $u\in(\text{cl } M-\text{cl } M)$.

Apparently, lemma (1.3) holds for C-semiconvex sets as well.

Choose linearly independent vectors $v^1, \ldots, v^m\in\text{lin } C$ satisfying (4.1). Let us consider the mapping (1.7) and define the functions $g_1, \ldots, g_m$ as follows:

$$\psi(x)=x+g_1(x)\,v^1+\ldots+g_m(x)\,v^m\,. \tag{4.3}$$

Lemma (4.4): *If M is C-semiconvex, then the functions $g_1, \ldots, g_m$ are convex.*

Proof: For any $x^1, x^2\in M+\text{lin } C$, $t\in[0, 1]$ and $x(t)=(1-t)\,x^1+tx^2$ we have

$\psi(x(t))=\psi(y(t))$, where

$$y(t)=(1-t)\,\psi(x^1)+t\psi(x^2)=x+\sum_i\left((1-t)\,g_i(x^1)+tg_i(x^2)\right)v^i\,.$$

$y(t)\in[M]\subset M+C$ yields $\psi(x(t))\in y(t)-C$ and hence

$$\psi(x(t))=y(t)-\sum_i\mu_i v^i,\quad \mu_i\geqq 0\,.$$

Consequently

$$g_i(x(t))\leqq(1-t)\,g_i(x^1)+tg_i(x^2),\quad i=1,\ldots,m$$

which has been to be proved. ▮

This lemma yields the following

Theorem (4.5): *Let M be a C-semiconvex set. Then ψ is continuous in* ri $(M+$ $+\operatorname{lin} C)$.

Definition (4.6): P is called a *C-semiconvex manifold* if it is C-semiconvex and $P+\operatorname{lin} C$ is a relatively open set.

Theorem (4.7): *A C-semiconvex manifold is homeomorphic to a relatively open convex set.*

Proof: Consider $P'=(P+L)\cap L^{\perp}$, where $L=\operatorname{lin} C$. The orthogonal projection $\pi\colon P\to P'$ is a homeomorphism according to theorem (4.5). ▮

We define $\dim P=\dim P'$.

Theorem (4.8): *Let P be a C-semiconvex manifold and $M\subset P$ be a compact C-semiconvex set. Then M is C-convex.*

Proof: The functions g_i are convex in $P+\operatorname{lin} C$ and therefore $\psi\colon(M+\operatorname{lin} C)\to M$ is a LIPSCHITZ type mapping. It means that M is C-regular and therefore it is C-convex. ▮

Let us consider a monotonic sequence of sets

$$M_1\subset M_2\subset\ldots\subset M_i\subset\ldots \tag{4.9}$$

Theorem (4.10): *The union $M=\bigcup_{i=1}^{\infty}M_i$ of a sequence* (4.9) *of C-semiconvex sets is C-semiconvex.*

Proof: For arbitrary $x^1, x^2\in M$, there exists an index i such that $x^1, x^2\in M_i$. Therefore, $(M-M)\cap\operatorname{lin} C=\{0\}$. According to the assumption, $[x^1, x^2]\subset M_i+$ $+C\subset M+C$. Hence $M^{[2]}\subset M+C$ and therefore $M+C$ is convex. ▮

Theorem (4.11): *Let C be a k-dimensional pointed closed convex cone. Then P is an $(n-k)$-dimensional C-semiconvex manifold if and only if the following two conditions are satisfied:*

1⁰. *$P+\operatorname{lin} C$ is an open set;*

2⁰. *For every couple $x^1, x^2\in P$, there exists a C-arc connecting x^1, x^2 in P.*

Proof: The condition 2^0 yields $(P-P)\cap \operatorname{lin} C=\{0\}$ and in addition, for every $x\in[x^1, x^2]$ we have $x\in\omega(x^1, x^2)+C$ so that $P^{[2]}\subset P+C$. Hence, P is C-semiconvex which together with 1^0 proves the "if" part of the theorem. The converse direction is trivial. ▮

Theorem (4.12): *The inclusion*

$$X\subset \operatorname{bd}[X] \tag{4.13}$$

holds for any subset of a C-semiconvex set M.

Proof: ${}^0C$ is a convex set because C is pointed. Thus $M+{}^0C=M+C+{}^0C$ is also convex and therefore $\operatorname{int}[M]\subset[M]+{}^0C=M+{}^0C$. Using $M\cap(M+{}^0C)=\emptyset$ we obtain $M\cap\operatorname{int}[M]=\emptyset$ and consequently $X\cap\operatorname{int}[X]=\emptyset$ for every $X\subset M$ which is equivalent to (4.13). ▮

V. Relatively Convex Functions

We shall discuss only finite real functions in R^n. The domain of definition of a given function f is denoted by dom f.

Definition (5.1): Let $C\subset R^n$ be a closed convex cone. A function defined in a subset of R^n is called *C-convex* if its epigraph is a C-convex subset of R^{n+1}.

Note (5.2): If the cone C in definition (5.1) is not specified, the function f is called relatively convex.

Lemma (5.3): *Let f be a C-convex function. Then*

$$F'=\{(x,\mu)\mid x\in\operatorname{dom} f,\ \mu\in R\}$$

is a C-convex set in R^{n+1}.

Proof: 1. Choose arbitrarily (x^1,μ^1), $(x^2,\mu^2)\in F'$. Then there exist numbers $\bar\mu^1,\bar\mu^2$ such that $\bar\mu^2-\bar\mu^1=\mu^2-\mu^1$ and $\bar\mu^1\geqq f(x^1)$, $\bar\mu^2\geqq f(x^2)$. Thus $(x^1,\bar\mu^1)$, $(x^2,\bar\mu^2)\in\operatorname{epi} f$. It means that $F'-F'=\operatorname{epi} f-\operatorname{epi} f$ which yields the regularity of F' with respect to C.

2. Consider $L'=\operatorname{lin} e^{n+1}$. Then $F'=L'+\operatorname{epi} f$ so that the sum $F'+C=L'+(\operatorname{epi} f+C)$ is convex by the hypothesis. ▮

Theorem (5.4): *The domain of definition of a C-convex function is C-convex.*

Proof: Consider a C-convex function $f\colon M\to R$. Since $C\subset R^n$ the space R^n may be held for an L-cylinder for $L=\operatorname{lin} C$. Using lemma (5.3) and theorem (2.28) we obtain that $M=F'\cap R^n$ is C-convex. ▮

Theorem (5.5): *Let f be a C-convex function. Then*

$$\operatorname{lev}_\alpha f=\{x\mid f(x)\leqq\alpha,\ x\in\operatorname{dom} f\}$$

is a C-convex set for any $\alpha\in R$.

Proof: Let $L'=\operatorname{lin} e^{n+1}$, $L=\operatorname{lin} C$. Choose $\alpha\in R$ and consider the sets

$$P_\alpha=(\alpha e^{n+1}+R^n)\cap\operatorname{epi} f,\quad Q_\alpha=P_\alpha+L'.$$

The hyperplane $W_\alpha = \alpha e^{n+1} + R^n$ is an L-cylinder which yields the C-convexity of P_α according to theorem (2.28). Hence Q_α is also C-convex by theorem (2.10). The proof is completed by applying again theorem (2.28) to the set $\operatorname{lev}_\alpha f = Q_\alpha \cap R^n$. ∎

Consider a C-convex set M and $x^1, x^2 \in M$. It has been already shown that the C-arc $\omega_{MC}(x^1, x^2)$ in M can be given using the parametric expression

$$\omega_{MC}(t; x^1, x^2) = \psi(x(t)) \quad \text{for} \quad x(t) = (1-t)\, x^1 + t x^2, \quad 0 \leqq t \leqq 1,$$

where

$$\psi(x) = (x + \operatorname{lin} C) \cap M = (x - C) \cap M \quad \text{for} \quad x \in [M] .$$

We shall show now that the concept of C-convex function can be defined using a relation linking the value of the function in the points of any C-arc.

Theorem (5.6): *Let M be a C-convex set. Then a function $f\colon M \to R$ is C-convex if and only if*

$$f(\omega_{MC}(t; x^1, x^2)) \leqq (1-t)\, f(x^1) + t f(x^2) \tag{5.7}$$

for every $x^1, x^2 \in M$ and $t \in [0, 1]$.

Proof: Consider the set

$$\bar{M} = \{(x, \mu) \mid x \in M,\ \mu \in R\} \subset R^{n+1} .$$

If $x^1, x^2 \in M$, then the points $\bar{x}^1 = (x^1, \alpha)$, $\bar{x}^2 = (x^2, \beta)$ belong to $\bar{M}$ for any $\alpha, \beta \in R$. Put $x(t) = (1-t)\, x^1 + t x^2$, $\bar{x}(t) = (1-t)\, \bar{x}^1 + t \bar{x}^2$ and $\mu(t) = (1-t)\, \alpha + t\beta$. The parallel projections with the direction $L = \operatorname{lin} C$ into the sets M, $\bar{M}$ are denoted by ψ and $\bar{\psi}$ respectively. Clearly

$$\bar{\psi}(\bar{x}(t)) = (\psi(x(t)), \mu(t)) \quad \text{for} \quad t \in [0, 1] .$$

1. Let f be a C-convex function. Then for every $\bar{x}^1, \bar{x}^2 \in \operatorname{epi} f$ we have $\bar{\psi}(\bar{x}(t)) \in \operatorname{epi} f$ and consequently

$$f(\omega_{MC}(t; x^1, x^2)) \leqq (1-t)\, \alpha + t\beta \quad \forall \alpha \geqq f(x^1), \quad \beta \geqq f(x^2) \tag{5.8}$$

which yields (5.7).

2. Let (5.7) holds. Then (5.8) holds as well which means that for every $\bar{x}^1$, $\bar{x}^2 \in \operatorname{epi} f$ there exists the C-arc connecting $\bar{x}^1$, $\bar{x}^2$. In addition, M, $\bar{M}$ are C-regular and hence $\operatorname{epi} f \subset \bar{M}$ is also C-regular. Thus $\operatorname{epi} f$ is a C-convex set according to theorem (3.5). ∎

Theorem (5.9): *Let $f\colon M \to R$ be a relatively convex function. Then there exist a convex set $\tilde{M}$ and a homeomorphism $\varphi\colon M \to \tilde{M}$ such that the composed function $\tilde{f} = f(\varphi^{-1})$ is convex in $\tilde{M}$.*

Proof: There exists a subspace L such that f is L-convex. Then M is an L-convex set by theorem (5.4). Consider the orthogonal projection $\varphi\colon M \to \tilde{M}$ of M into $L^\perp$. Then $\tilde{M} = (M + L) \cap L^\perp$ is a convex set and φ is a homeomorphism according to theorem (1.17). Choose $\tilde{x}^1, \tilde{x}^2 \in \tilde{M}$ and let $x^i = \varphi^{-1}(\tilde{x}^i)$, $i = 1, 2$. Then the C-arc $\omega_C(t; x^1, x^2)$ can be expressed as $\varphi^{-1}((1-t)\, \tilde{x}^1 + t\tilde{x}^2)$, $t \in [0, 1]$. Theorem (5.6) yields

$$\tilde{f}\left((1-t)\,\tilde{x}^1+t\tilde{x}^2\right) \leqq (1-t)\,\tilde{f}(\tilde{x}^1)+t\tilde{f}(\tilde{x}^2)$$

for every $\tilde{x}^1, \tilde{x}^2 \in \tilde{M}$ and $t \in [0, 1]$. Thus $\tilde{f}$ is convex. ∎

The theorem (5.9) ensures that relatively convex functions keep the most useful properties of convex functions.

Corollary (5.10): *A relatively convex function f is continuous in* ri (dom f).

Proof: The function $\tilde{f}=f(\varphi^{-1})$ is convex in $\tilde{M}$ and therefore it is continuous in ri $\tilde{M}$. Hence $f=\tilde{f}(\varphi)$ is continuous in ri (dom f) = ri $M=\varphi^{-1}$ (ri $\tilde{M}$) since φ is homeomorphic. ∎

Corollary (5.11): *Every local minimum of a relatively convex function f is a global one.*

Proof: Let x^1, x^2 be minimum points of f. Then $\tilde{x}^1, \tilde{x}^2$ are minimum points of $\tilde{f}$. If $f(x^1)<f(x^2)$ held then it would be $\tilde{f}(\tilde{x}^1)<\tilde{f}(\tilde{x}^2)$ which would contradict the convexity of $\tilde{f}$. Thus $f(x^1)=f(x^2)$. ∎

To utilize the concept of a relatively convex function we need find some practically available criteria of relative convexity. A simple problem of this kind may be formulated as follows:

There are an open set G, a C-convex function $M \subset G$ and a convex function $f: G \to R$. It is to find out whether the restricted function $f_M: M \to R$ ($f_M(x)=f(x)$ for $x \in M$) is C-convex.

Let $\partial f(x)$ be the set of all subgradients of a convex function f at a point x and let us denote $\partial f(M) = \bigcup \{\partial f(x) \mid x \in M\}$.

Theorem (5.12): *Let G be an open convex set and $M \subset G$ be a C-convex set. Let a function $f: G \to R$ satisfy the conditions:*

1⁰. f is convex;

2⁰. $\partial f(M) \subset (-C)^*$.

Then the restricted function f_M is C-convex.

Proof: Choose $x^1, x^2 \in M$, $t \in [0, 1]$ and put $x(t)=(1-t)\,x^1+tx^2$, $y(t)=\omega_C(t; x^1, x^2)$ and finally $u=x(t)-y(t)$. Clearly $u \in C$.
According to 2⁰ $u \in (-\partial f(y(t)))^*$ which implies

$$f(y(t)) \leqq f(x(t))\,. \tag{5.13}$$

Due to the convexity of f

$$f(x(t)) \leqq (1-t)\,f(x^1)+t\,f(x^2)\,,$$

which together with (5.13) yields

$$f_M(\omega_C(t; x^1, x^2)) = f_M(y(t)) \leqq (1-t)\,f_M(x^1)+tf_M(x^2)\,.$$

Thus, f_M is C-convex according to theorem (5.6). ∎

References

[1] ARROW, K. J., L. HURWICZ and H. UZAWA: Studies in Linear and Nonlinear Programming. Stanford, Stanford University Press 1958.
[2] KUHN, H. W., and A. W. TUCKER: Linear Inequalities and Related Problems. Princeton, Princeton University Press 1956.
[3] MARTOS, B.: Nonlinear Programming–Theory and Methods. Budapest, Akadémiai Kiadó 1973.
[4] NEDOMA, J.: Relatively Convex Analysis. Research publ. No. 50, EML EÚ ČSAV, Praha 1977.
[5] ROCKAFELLAR, R. T.: Vypuklyj analiz. Moscow, Mir 1973 (translated from English).
[6] STOER, J., and C. WITZGALL: Convexity and Optimization in Finite Dimensions I. Berlin, Springer 1970.
[7] YU, P. L.: Cone Convexity, Cone Extreme Points and Nondominated Solutions in Decision Problems with Multiobjectives. Journal of Optimization Theory and Application 14 (1974) 3.

Zusammenfassung

Um das Konzept einer konvexen Menge auf Untermengen von linearen Mannigfaltigkeiten anzuwenden, wird die sogenannte relative Konvexität bezüglich eines konvexen Kegels definiert. Dieser Begriff der relativen Konvexität liefert die Möglichkeit, das Konzept einer konvexen Funktion so zu verallgemeinern, daß es zur Formulierung globaler Optimalitätskriterien für eine Klasse nichtkonvexer Optimierungsprobleme benutzt werden kann.

Резюме

Для применения концепта выпуклого множества к подмножествам линейных многообразий определяется так называемая относительная выпуклость по отношению к выпуклому конусу. Относительная выпуклость предоставляет возможность, обобщать концепт выпуклой функции таким образом, что он может быть использован к формулировке глобальных критериев для некоторого класса невыпуклых проблем оптимизации.

Eingereicht bei der Redaktion: 21. 12. 1977; revidierte Fassung: 22. 5. 1978

Math. Operationsforsch. Statist., Ser. Optimization, Vol. 10 (1979) No. 1, 17–25

Linearität und Verallgemeinerungen der Konvexität[1]

KAREL ZIMMERMANN[2]

Zusammenfassung: Es wird versucht, mit Hilfe einer Axiomatik der verallgemeinerten Konvexität bis zu einer axiomatischen Einführung von Hyperebenen zu kommen, den Begriff der Linearität einzuführen und einige Eigenschaften der in diesem Sinne linearen Optimierungsaufgaben zu untersuchen. Das sind z. B. solche Eigenschaften, die bei der numerischen Lösung von bestimmten Klassen der Optimierungsaufgaben ausgenutzt werden. Diese Eigenschaften liefern auch eine Methodologie zur Aufstellung von Algorithmen für die Lösung weiterer Optimierungsaufgaben dieser Art.

Verschiedene Arbeiten (z. B. [1], [2], [3]) befassen sich mit Verallgemeinerungen der Konvexität. Im Zusammenhang mit dieser Problematik und ihrer eventuellen Anwendung bei der Lösung von Optimierungsaufgaben entsteht die Frage, was bedeutet die Linearität in einem Raum mit einem verallgemeinerten Konvexitätsbegriff. Diese Frage ist dann sehr eng mit der Definition von Hyperebenen in solchen Räumen verbunden. Mit Hilfe einer Axiomatik der verallgemeinerten Konvexität wird versucht, bis zu einer axiomatischen Einführung von Hyperebenen zu kommen, den Begriff der Linearität einzuführen und einige Eigenschaften der in diesem Sinne linearen Optimierungsaufgaben zu untersuchen. Diese Eigenschaften wurden bei der numerischen Lösung von bestimmten Klassen der Optimierungsaufgaben benutzt. Die Eigenschaften geben uns auch eine Methodologie an, wie Algorithmen für die Lösung von weiteren Optimierungsaufgaben dieser Art zusammengestellt werden können. Der Vortrag benutzt Aussagen, die in [4] bewiesen wurden.

Definition 1: Wir sagen, daß auf $\boldsymbol{P}$ eine *Verbindungsoperation* definiert ist, wenn jedem Paar $a, b \in \boldsymbol{P}$ eine Untermenge $(ab) \subset \boldsymbol{P}$ zugeordnet ist.

Wir definieren die Mengen (AB), a/b, $\text{Kern}_R(M)$, Kern (M) für $A, B, M, R \subset \boldsymbol{P}$; $a, b \in \boldsymbol{P}$:

$$(AB) \equiv \bigcup_{\substack{a \in A \\ b \in B}} (ab);$$

$$a/b \equiv \{z \mid z \in \boldsymbol{P},\ a \in (zb)\};$$

[1] Vortrag, gehalten auf der Tagung „Mathematische Optimierung – Theorie und Anwendungen" in Eisenach (DDR), 5.–10. Dezember 1977.

[2] Karls-Universität Prag, Abt. Kybernetik und Operationsforschung, Malostr. nám 25, ČS – 11800 Praha 1, ČSSR.

$\text{Kern}_R(M) \equiv \{a \mid a \in M$, für alle $r \in R$ existieren $b \in (ar) \cap (M \setminus \{a\})$, $c \in M \setminus \{a\}$, so daß $a \in (bc)\}$;

$\text{Kern}(M) \equiv \text{Kern}_P(M)$.

Definition 2:

(1) Die Menge $M \subset \boldsymbol{P}$ heißt *konvex*, wenn $a, b \in M \Rightarrow (ab) \subset M$ gilt.

(2) Das Element $c \in M$ heißt ein *Eckpunkt* von M, wenn gilt:
$[a, b \in M, c \in (ab)] \Rightarrow$[entweder $c = a$ oder $c = b$] .

Definition 3: Die Menge $\boldsymbol{P}$ mit einer Verbindungsoperation (ab) für $a, b \in \boldsymbol{P}$ heißt ein *Konnexitätsraum*, wenn folgende fünf Axiome erfüllt sind:

(A1) $a \in (ab)$, $b \in (ab)$;
(A2) $(ab) = (ba)$;
(A3) $(ab)\,c = a(bc)$;
(A4) a/b ist konvex;
(A5) $\text{Kern}(\boldsymbol{P}) \neq \emptyset$.

Definition 4: Sei $H \subset \boldsymbol{P}$, $H \neq \emptyset$. Die Menge H heißt *Hyperebene* in $\boldsymbol{P}$, wenn dieser Menge zwei konvexe Mengen $H^{(1)}(H)$, $H^{(2)}(H)$ mit folgenden fünf Eigenschaften zugeordnet werden können:

(E1) $H^{(1)}(H) \cap H^{(2)}(H) = H$;
(E2) $H^{(1)}(H) \cup H^{(2)}(H) = \boldsymbol{P}$;
(E3) $\text{Kern}(H^{(i)}(H) \setminus H) = H^{(i)}(H) \setminus H \neq \emptyset$ für $i = 1, 2$;
(E4) $[a \in H^{(i)}(H) \setminus H,\ b \in H^{(j)}(H)] \Rightarrow H \cap (ab) \neq \emptyset$ für $i \neq j$;
(E5) $[a \in H^{(1)}(H) \setminus H$ bzw. $H^{(2)}(H) \setminus H$, $c \in (ab) \cap H] \Rightarrow b \in H^{(2)}(H)$ bzw. $H^{(1)}(H)$.

Die Mengen $H^{(1)}(H)$, $H^{(2)}(H)$ werden wir im weiteren die zu der Hyperebene H zugehörigen *Halbräume* nennen.

Lemma 1: *Es sei H eine Hyperebene in $\boldsymbol{P}$. Dann sind die Mengen $H^{(1)}(H) \setminus H$, $H^{(2)}(H) \setminus H$ konvex.*

Beweis: Wir setzen voraus, daß z. B. die Menge $H^{(1)}(H) \setminus H$ nicht konvex ist. Dann existieren Elemente $a, b \in H^{(1)}(H) \setminus H$, $c \notin H^{(1)}(H) \setminus H$, so daß $c \in (ab)$. Es ist dann entweder $c \in H$ oder $c \in H^{(2)}(H) \setminus H$. Der Fall $c \in H$ steht aber im Widerspruch zu (E5) aus Definition 4. Ist jetzt $c \in H^{(2)}(H) \setminus H$, existiert nach (E4) aus Definition 4 ein Element $\bar{c} \in (ac) \cap H$. Dann haben wir aber $a, b \in H^{(1)}(H) \setminus H$, $(ac) \subset (ab)$[3], so daß $\bar{c} \in (ab)$, $\bar{c} \in H$, was wieder im Widerspruch zu (E5) aus Definition 4 steht. Es muß also immer $(ab) \subset H^{(1)}(H) \setminus H$ sein. Analog kann man die Konvexität der Menge $H^{(2)}(H) \setminus H$ beweisen. ∎

Definition 5: Sei $\{A_\alpha\}_{\alpha \in I}$ ein System von Untermengen in $\boldsymbol{P}$, $L = \bigcap_{\alpha \in I} A_\alpha$, $I' \subset I$. Das System $\{A_\alpha\}_{\alpha \in I'}$ heißt *minimales Subsystem* des Systems $\{A_\alpha\}_{\alpha \in I}$,

[3] Es ist nämlich $a \in (ab)$, $c \in (ab)$, und die Menge (ab) ist nach Lemma 7 konvex, wie im Anhang bewiesen wird.

wenn $\bigcap_{\alpha\in I'} A_\alpha = L$ und $\bigcap_{\alpha\in I'\setminus\{\beta\}} A_\alpha \neq L$ für alle $\beta\in I'$. Das System $\{A_\alpha\}_{\alpha\in I}$ heißt *minimal*, wenn gilt:

$$\bigcap_{\alpha\in I\setminus\{\beta\}} A_\alpha \neq L \quad \text{für alle} \quad \beta\in I\,.$$

Definition 6: Ein System $\{H_\alpha\}_{\alpha\in I^{(1)}}$ von Hyperebenen in $\boldsymbol{P}$ heißt *geometrisch unabhängig*, wenn dieses System zu einem System von Hyperebenen $\{H_\alpha\}_{\alpha\in I}$, $I^{(1)}\subset I$, so vervollständigt werden kann, daß das System $\{H_\alpha\}_{\alpha\in I}$ minimal ist und die Menge $\bigcap_{\alpha\in I} H_\alpha$ nur ein Element besitzt.

Definition 7: Der Konnexitätsraum $\boldsymbol{P}$ hat die Dimension n ($\dim \boldsymbol{P}=n$), wenn folgendes gilt:

(1) Es existiert ein geometrisch unabhängiges System von Hyperebenen in $\boldsymbol{P}$.

(2) Es sei $\{H_\alpha\}_{\alpha\in I}$ ein beliebiges geometrisch unabhängiges System von Hyperebenen in $\boldsymbol{P}$, $L=\bigcap_{\alpha\in I} H_\alpha$. Dann gilt:

(a) $|L|=1 \Rightarrow |I|=n$

(b) $\mathrm{Kern}_L(L)=L\cap \mathrm{Kern}\,(\boldsymbol{P})$.

Folgender Satz kann jetzt bewiesen werden:

Satz 1: *Es sei* Kern $(\boldsymbol{P})=\boldsymbol{P}$, $\dim \boldsymbol{P}=n$, *und es sei* $\{H_1, \ldots, H_k\}$ *ein solches System von Hyperebenen in* $\boldsymbol{P}$, *daß jedes Subsystem dieses Systems mit n Hyperebenen mit nichtleerem Durchschnitt geometrisch unabhängig ist. Seien* $P_1, \ldots, P_k$ *die zu* $H_1, \ldots, H_k$ *zugehörigen Halbräume,* $L=\bigcap_{r=1}^{k} P_r$. *Dann ist die Menge der Eckpunkte von L entweder leer oder endlich.*

Die folgenden Lemmata dienen zum Beweis des Satzes 1.

Lemma 2: *Seien* A_i, $i=1, \ldots, k$, *konvexe Mengen,* $L=\bigcap_{i=1}^{k} A_i \neq \emptyset$. *Die Menge L ist konvex.*

Beweis: Seien $a, b\in L$, d. h. $a, b\in A_i$ für alle i. Es folgt aus der vorausgesetzten Konvexität der Mengen A_i:

$$a, b\in A_i \Rightarrow (ab)\subset A_i \quad \text{für alle} \quad i\,.$$

Das heißt aber, daß $(ab)\subset L$, und die Menge L ist konvex. ∎

Bemerkung: Aus dem Lemma 2 folgt unmittelbar wegen (E1) die Konvexität jeder Hyperebene in $\boldsymbol{P}$.

Weiter bezeichnen wir in diesem Artikel mit $K(U)$ die Menge von allen Eckpunkten einer beliebigen Menge U, $U\subset \boldsymbol{P}$.

Lemma 3: *Seien* A, H *konvexe Mengen in* $\boldsymbol{P}$, *deren Durchschnitt mehr als ein Element besitzt. Sei* $a\in \mathrm{Kern}\,(H)\cap \mathrm{Kern}_A\,(A)$. *Es gilt:* $a\notin K\,(A\cap H)$.

Beweis: $a\in \mathrm{Kern}\,(H)\cap \mathrm{Kern}_A\,(A) \Rightarrow a\in A\cap H$. Sei $b'\in A\cap H, b'\neq a$. Da $a\in$ $\in \mathrm{Kern}_A\,(A)$, existierten für dieses Element b' ein Element $b\in (b'a)\cap (A\setminus\{a\})$ und ein

Element $c \in A$, $c \neq a$, so daß $a \in (bc)$. Da die Mengen A, H konvex sind, ist auch die Menge $A \cap H$ konvex (Lemma 2), und es ist folglich $(b'a) \subset A \cap H$ und $b \in A \cap H$ (da $b \in (b'a)$). Da $a \in \text{Kern}\,(H)$ ist, existieren für das Element $c \in A \subset \boldsymbol{P}$ Elemente $d \in (ac) \cap (H \setminus \{a\})$ und $f \in H \setminus \{a\}$, so daß $a \in (df)$. Wir untersuchen jetzt die Eigenschaften des Elements d. Es ist a, $c \in a/b$, und die Menge a/b ist nach (A4) konvex; infolgedessen ist $(ac) \subset a/b$ und deshalb auch $d \in a/b$. Da die Menge A konvex ist und a, $c \in A$, gilt $(ac) \subset A$ und deshalb auch $d \in A$ (da $d \in (ac)$). Es folgt also aus der Konstruktion des Elements d: $d \in A \cap H \cap (a/b)$, $d \neq a$. Aus der Definition der Menge a/b erhalten wir, daß $a \in (bd)$; dabei ist b, $d \in (A \cap H) \setminus \{a\}$, so daß $a \notin K\,(A \cap \cap H)$. ▮

Lemma 4: *Seien $A_1, \ldots, A_k$ konvexe Mengen in $\boldsymbol{P}$ mit der Eigenschaft* Kern $(A_i) = A_i$ *für $i = 1, \ldots, k$. Sei $L = \bigcap_{i=1}^{k} A_i \neq \emptyset$. Dann gilt* Kern $(L) = L$.

Beweis: Aus der Definition der Menge Kern (L) folgt Kern $(L) \subset L$. Es bleibt also zu beweisen, daß $L \subset \text{Kern}\,(L)$, d. h. $a \in L \Rightarrow a \in \text{Kern}\,(L)$. Sei $a \in L$ beliebig. Es ist dann $a \in A_i = \text{Kern}\,(A_i)$ für alle i. Sei $b \in \boldsymbol{P}$, $b \neq a$, beliebig. Da $a \in \text{Kern}\,(A_1)$, existieren Elemente $c_1 \in (ab) \cap (A_1 \setminus \{a\})$ und $d_1 \in A_1 \setminus \{a\}$, so daß $a \in (d_1 c_1)$. Da $a \in \in \text{Kern}\,(A_2)$, existiert für das Element $d_1 \in \boldsymbol{P}$ ein Element $d_2 \in (d_1 a) \cap (A_2 \setminus \{a\})$ (die Existenz dieses Elements d_2 folgt aus der Definition der Menge Kern (A_2)). Es ist dabei a, $d_1 \in a/c_1$, und die Menge a/c_1 ist nach (A4) konvex. Es ist also $(d_1 a) \subset a/c_1$ und deshalb $d_2 \in a/c_1$. Da d_1, $a \in A_1$ und die Menge A_1 nach Voraussetzung konvex ist, gilt auch $(d_1 a) \subset A_1$ und deshalb auch $d_2 \in A_1$. Es ist also $d_2 \in A_1 \cap A_2 \cap \cap (a/c_1)$, $d_2 \neq a$. Aus der Definition der Menge a/c_1 erhalten wir $a \in (d_2 c_1)$. Da $a \in \text{Kern}\,(A_2)$, existiert für das Element c_1 ein Element $c_2 \in (ac_1) \cap (A_2 \setminus \{a\})$; da a, $c_1 \in a/d_2$ und die Menge a/d_2 konvex ist, erhalten wir, daß $c_2 \in a/d_2$ und deshalb $a \in (d_2 c_2)$, wobei $c_2 \in A_2$, $c_2 \in (ac_1) \subset A_1$ (da a, $c_1 \in A_1$ und A_1 konvex ist). Es ist also $a \in (d_2 c_2)$, wobei $d_2 \in A_1 \cap A_2$, $c_2 \in A_1 \cap A_2$ und $d_2 \neq a$, $c_2 \neq a$. Da a, $c_1 \in (ab)$ und die Menge (ab) konvex ist (vgl. Anhang, Lemma 7), erhalten wir noch $c_2 \in (ac_1) \subset \subset (ab)$. Wenn wir diese Konstruktion fortsetzen, erhalten wir nach k Schritten Elemente d_k, $c_k \in \bigcap_{i=1}^{k} A_i \setminus \{a\}$, $c_k \in (ab)$ so, daß $a \in (d_k c_k)$. Das heißt aber, daß $a \in \in \text{Kern}\,(L)$ (für beliebiges $b \in L$, $b \neq a$ haben wir Elemente $c_k \in (ab) \cap L \setminus \{a\}$ und $d_k \in L \setminus \{a\}$ mit der Eigenschaft $a \in (c_k d_k)$ gefunden). Es ist also $L \subset \text{Kern}\,(L)$ und deshalb $L = \text{Kern}\,(L)$. ▮

Beweis des Satzes 1: Ist $L = \emptyset$, so ist auch die Menge $K(L)$ leer. Seien jetzt $L \neq \emptyset$, $a \in L$. Wir setzen $P_i^0 \equiv P_i \setminus H_i$ für alle i. Nach (E3) aus der Definition 4 gilt Kern $(P_i^0) = P_i^0$ für alle i. Wir zeigen jetzt, daß unter den gemachten Voraussetzungen das Element $a \in L$ nur dann in $K(L)$ liegen kann, wenn es im Durchschnitt von genau n Hyperebenen von $H_1, \ldots, H_k$ enthalten ist. Mit anderen Worten, wir zeigen, daß, wenn a im Durchschnitt von weniger oder mehr als n dieser Hyperebenen liegt, a nicht in $K(L)$ enthalten sein kann. Daraus ergibt sich schon die Endlichkeit der Menge $K(L)$ (die Anzahl von verschiedenen Kombinationen der n Hyperebenen aus $\{H_1, \ldots, H_k\}$ ist endlich). Wir werden weiter folgende drei Fälle unterscheiden:

Fall 1: $a \in \bigcap_{i=1}^{k} P_i^0$.

Fall 2: $a \in (\bigcap_{i \in I} H_i) \cap (\bigcap_{i \notin I} P_i^0)$, $|I| \leqq n-1$.

Fall 3: $a \in (\bigcap_{i \in I} H_i) \cap (\bigcap_{i \notin I} P_i^0)$, $|I| > n$.

Wir zeigen, daß in jedem der drei Fälle a nicht in der Menge $K(L)$ enthalten ist.

Fall 1: Nach dem Lemma 4 und (E3) aus der Definition 4 haben wir in diesem Fall

$$a \in \bigcap_{i=1}^{k} P_i^0 = \bigcap_{i=1}^{k} \text{Kern } (P_i^0) = \text{Kern}\left(\bigcap_{i=1}^{k} P_i^0\right).$$

D. h., für beliebiges $b \in \boldsymbol{P}$, $b \neq a$, existieren Elemente $c, d \in \bigcap_{i=1}^{k} P_i^0 \subset L$, $c \in (ab)$, $c \neq a$, $d \neq a$, so daß $a \in (dc)$ und deshalb $a \notin K(L)$.

Fall 2: Setzen wir $\tilde{L} \equiv \bigcap_{i \in I} H_i$, $T \equiv \bigcap_{i \notin I} P_i^0$. Aus der Voraussetzung Kern $\boldsymbol{P} = \boldsymbol{P}$ ergibt sich: $\tilde{L} \cap \text{Kern } (\boldsymbol{P}) = \tilde{L} \cap \boldsymbol{P} = \tilde{L}$. Es ist weiter $|\tilde{L}| \neq 1$ (wäre $|\tilde{L}| = 1$, so müßte $|I| = n$ sein, vgl. Definition 7–(2) (a)). Nach der Definition 7–(2) (b) erhalten wir: $\text{Kern}_{\tilde{L}} (\tilde{L}) = \tilde{L} \cap \text{Kern } (\boldsymbol{P}) = \tilde{L} \cap \boldsymbol{P} = \tilde{L}$. Aus (E3) (Definition 4) und Lemma 4 erhalten wir weiter:

$$\text{Kern } (T) = T .$$

Weiter ist: $a \in T \cap \tilde{L} = \text{Kern } (T) \cap \text{Kern}_{\tilde{L}} (\tilde{L})$. Da $|\tilde{L}| \neq 1$ ist, existiert ein Element $f \in \tilde{L}$, $f \neq a$. Da $a \in \text{Kern } (T)$, existiert für dieses Element f ein Element $p \in (af) \cap \cap (T \setminus \{a\})$ (das folgt aus der Definition der Menge Kern (T)). Aus der Konvexität der Menge $\tilde{L}$ (vgl. Bemerkung und Lemma 2) und aus $a, f \in \tilde{L}$ ergibt sich $(af) \subset \tilde{L}$, so daß $p \in T \cap \tilde{L}$, $p \neq a$. Das bedeutet, daß die Menge $T \cap \tilde{L}$ mehr als ein Element besitzt und alle Voraussetzungen aus dem Lemma 3 erfüllt sind (T ist konvex laut Definition 4 und Lemma 2). Nach diesem Lemma erhalten wir, daß $a \notin K(T \cap \cap \tilde{L})$ ist; d. h., es existieren Elemente $d, c \in T \cap \tilde{L} \subset L$, $d \neq a$, $c \neq a$, so daß $a \in (dc)$ und deshalb auch $a \notin K(L)$ gilt.

Fall 3: In diesem Fall ist $|I| > n$. Setzen wir wieder $\tilde{L} \equiv \bigcap_{i \in I} H_i$, $T \equiv \bigcap_{i \notin I} P_i^0$. Aus der Voraussetzung über die geometrische Unabhängigkeit und aus der Definition 7–(2) (a) folgt $|\tilde{L}| \neq 1$. Wäre $|\tilde{L}| < 1$, hätten wir $\tilde{L} = \emptyset$, und die Menge $T \cap \tilde{L}$ könnte kein Element enthalten. Sei jetzt $|\tilde{L}| > 1$. Sei $I_1 \subset I$, $|I_1| = n$, $\tilde{L}_1 = \bigcap_{i \in I_1} H_i$. Da $a \in \tilde{L}_1$ ist, ist die Menge $\tilde{L}_1$ nichtleer. Nach Voraussetzung müssen die Hyperebenen $\{H_\alpha\}_{\alpha \in I_1}$ geometrisch unabhängig sein, und d. h. $|\tilde{L}_1| = 1$. Da $\tilde{L} \subset \tilde{L}_1$ ist, muß $|\tilde{L}| \leqq |\tilde{L}_1| = 1$ sein. Das steht aber im Widerspruch zu der vorausgesetzten Beziehung $|\tilde{L}| > 1$. Im Fall 3 ist also immer $\tilde{L} = \emptyset$, und $a \in \tilde{L}$ kann nicht existieren. ∎

Es sei jetzt $\boldsymbol{R}$ die Menge der reellen Zahlen.

Definition 8: Es sei $F: \boldsymbol{P} \to \boldsymbol{R}$ eine Funktion über $\boldsymbol{P}$, $k \in \boldsymbol{R}$ beliebig, $T_1^k = \{a \in \boldsymbol{P} \mid f(a) \leqq k\}$, $T_2^k = \{a \in \boldsymbol{P} \mid f(a) \geqq k\}$, $T^k = \{a \in \boldsymbol{P} \mid f(a) = k\}$. Die Funktion f heißt *geometrisch linear*, wenn gilt:

$T^k \neq \emptyset \Rightarrow T^k$ ist eine Hyperebene in $\boldsymbol{P}$, und T_1^k, T_2^k sind die zugehörigen Halbräume.

Eine *geometrisch lineare Optimierungsaufgabe* kann man jetzt folgendermaßen formulieren:

Man maximiere bzw. minimiere eine geometrisch lineare Funktion f über der Menge $L = \bigcap_{r=1}^{k} P_r$, *wobei* $P_1, \ldots, P_k$ *gegebene Halbräume in* $\boldsymbol{P}$ *sind.*

Definition 9: Sei $M \subset \boldsymbol{P}$. Die Hyperebene H heißt *Stützhyperebene* der Menge M, wenn folgendes gilt:

(1) $M \subset H^{(1)}(H)$ bzw. $H^{(2)}(H)$;

(2) $M \cap H \neq \emptyset$.

Satz 2: *Es sei eine geometrisch lineare Optimierungsaufgabe von oben angeführter Form mit* $K(L) \neq \emptyset$ *gegeben. Es sei weiter für jede Stützhyperebene* H *der Menge* L *die Implikation*

$$K(L) \neq \emptyset \Rightarrow K(L) \cap (L \cap H) \neq \emptyset$$

erfüllt. Dann existiert mindestens ein Eckpunkt von L, *in dem der Optimalwert der Funktion f angenommen wird.*

Beweis: Wir setzen voraus, daß die geometrisch lineare Optimierungsaufgabe eine Minimierungsaufgabe ist. Sei $\alpha = \min_{a \in L} f(a) = f(\bar{a})$. Die Menge $H_\alpha \equiv \{a \in \boldsymbol{P} \mid f(a) = \alpha\}$ ist dann eine Stützhyperebene der Menge L, da folgendes gilt:

(1) $L \subset T_2 = \{a \in \boldsymbol{P} \mid f(a) \geqq \alpha\}$;

(2) $H_\alpha \cap L \neq \emptyset$ (da $\bar{a} \in H_\alpha \cap L$ ist).

Die Menge $K(L)$ ist nichtleer, und deshalb muß nach der Voraussetzung auch die Menge $K(L) \cap (L \cap H_\alpha)$ nichtleer sein. Jeder Punkt dieser Menge ist dann ein gesuchter Eckpunkt von L, in dem der Optimalwert der Funktion f angenommen wird. ▮

Die Sätze 1 und 2 können bei der Lösung der geometrisch linearen Optimierungsaufgaben benutzt werden. Sind die Voraussetzungen dieser Sätze erfüllt, so ist die Menge $K(L)$ endlich, und mindestens eine optimale Lösung der geometrisch linearen Optimierungsaufgabe muß in $K(L)$ liegen. Um die Aufgabe zu lösen, muß man nur Punkte von $K(L)$ überprüfen. Unter Ausnutzung einiger spezieller Eigenschaften der konkreten Optimierungsaufgaben gelingt es manchmal, effektivere Algorithmen zu finden, bei denen nicht alle Punkte von $K(L)$ überprüft werden müssen. Das gilt z. B. für die gewöhnlichen linearen Optimierungsaufgaben und den Simplexalgorithmus oder für die extremal linearen Optimierungsaufgaben, die in [4] beschrieben werden. Wir bringen hier noch eine andere Aufgabe dieser Art, die auch in [4] angeführt wurde.

Beispiel 1: Sei $\boldsymbol{P}=\boldsymbol{R}_+\times\boldsymbol{R}_+$, wobei $\boldsymbol{R}_+=\{r\in\boldsymbol{R}\mid r\geqq 0\}$;

$$(XY)=\{Z\mid Z_1=\max(\lambda_1x_1,\lambda_2y_1),\, Z_2=\max(\lambda_1^2y_2,\lambda_2^2x_2),$$
$$\max(\lambda_1,\lambda_2)=1,\ \lambda_1,\lambda_2\in\boldsymbol{R}_+\}\,.$$

Die Menge L aus dem Satz 1 kann z. B. folgende Form besitzen (Abb. 1):

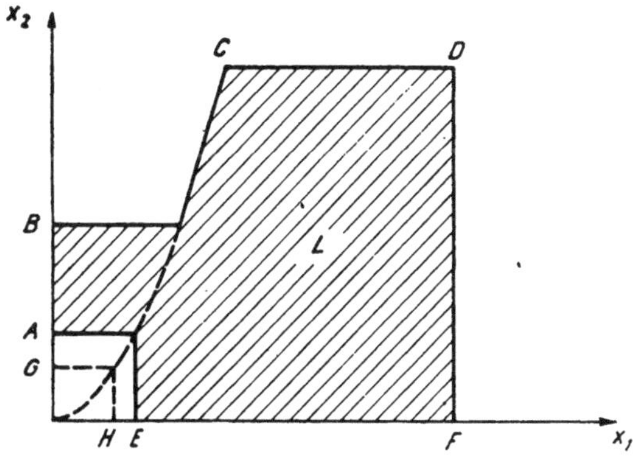

Abb. 1. GH ist eine Niveaulinie der Zielfunktion $f(x_1, x_2)=\max(x_1^2, x_2)$. $K(L)=\{A, B, C, D, E, F\}$.

Anhang

Wir zeigen, daß die Menge (ab) für jedes Paar a, b konvex ist. Diese Behauptung wurde im Beweis von Lemma 1 und 4 benutzt.

Lemma 5 (vgl. [2]): *Seien $A\subset B\subset\boldsymbol{P}$, $C\subset D\subset\boldsymbol{P}$. Dann gilt $(AC)\subset(BD)$.*

Beweis: Es sei $x\in(AC)$. Dann ist $x\in(ac)$, $a\in A$, $c\in C$. Da $A\subset B$, $C\subset D$, ergibt sich $a\in B$, $c\in D$ und deshalb $x\in(BD)$ und $(AC)\subset(BD)$. ∎

Lemma 6 (vgl. [2]): *Die Menge $A\subset\boldsymbol{P}$ ist dann und nur dann konvex, wenn gilt: $(AA)\subset A$*

Beweis: Sei A konvex, $x\in(AA)$, d. h. $x\in(ab)$, wobei $a, b\in A$ sind. Aus der Konvexität der Menge A erhalten wir: $x\in(ab)\subset A$ und deshalb $x\in A$, d. h. $(AA)\subset A$. Sei jetzt $(AA)\subset A$; $a, b\in A$. Dann haben wir: $(ab)\subset(AA)\subset A$, d. h., die Menge A ist konvex. ∎

Lemma 7 (vgl. [2]): *Seien $A, B\subset\boldsymbol{P}$, A, B konvex. Dann ist die Menge (AB) auch konvex.*

Beweis: Wegen (A3) (Definition 3) haben wir:
$(AB)\,(AB)=(AA)\,(BB)$. Nach dem Lemma 6 erhalten wir: $AA\subset A$, $BB\subset B$. Daraus folgt nach dem Lemma 5 $(AA)\,(BB)\subset(AB)$. Wir haben also: $(AB)\,(AB)=(AA)\,(BB)\subset$ $\subset(AB)$, und nach dem Lemma 6 erhalten wir die Konvexität der Menge (AB). ∎

Bemerkung: Für $A=\{a\}$, $B=\{b\}$ erhalten wir aus dem Lemma 7 die Konvexität der Menge $(AB)=(\{a\}\,\{b\})=(ab)$.

Lemma 8 (weitere Eigenschaften von Hyperebenen):
Es sei H eine Hyperebene in $\boldsymbol{P}$. Dann gilt:
(a) *H ist konvex;*
(b) $H^{(1)}(H)\neq H^{(2)}(H)$;
(c) *Das Paar $(H^{(1)}(H);\ H^{(2)}(H))$ ist bis auf die Reihenfolge der Komponenten eindeutig bestimmt.*

Beweis: Die Behauptung (a) folgt unmittelbar aus (E1) (Definition 4) und Lemma 2. Um die Behauptung (b) zu beweisen, setzen wir voraus, daß $H^{(1)}(H) = H^{(2)}(H)$ ist. Aus (E1) und (E2) ergibt sich dann: $H^{(1)}(H) = H^{(2)}(H) = H = \boldsymbol{P}$. Daraus folgt dann für $i = 1, 2$:

$$\text{Kern}\,(H^{(i)}(H) \backslash H) = H^{(i)}(H) \backslash H = H \backslash H = \emptyset\,,$$

was im Widerspruch zu (E3) (Kern $(H^{(i)}(H)\backslash H) = H^{(i)}(H)\backslash H \neq \emptyset$) steht. Für den Beweis der Behauptung (c) nehmen wir an, daß zwei verschiedene Paare

$$(H^{(1)}(H); H^{(2)}(H)) \quad \text{und} \quad (\bar{H}^{(1)}(H); \bar{H}^{(2)}(H))$$

mit den Eigenschaften aus Definition 4 der Hyperebene H zugeordnet werden können. Da sich die Paare nicht nur durch die Reihenfolge der Komponenten unterscheiden, kann nicht vorkommen, daß $H^{(1)}(H) = \bar{H}^{(2)}(H)$, $H^{(2)}(H) = \bar{H}^{(1)}(H)$. Ohne Beschränkung der Allgemeinheit kann man dann (nach einer eventuellen Umnumerierung der Komponenten) voraussetzen, daß zwei Punkte a, b mit folgenden Eigenschaften existieren:

$$a \in H^{(1)}(H)\backslash H, \quad a \notin \bar{H}^{(1)}(H)\backslash H \tag{V 1}$$

$$b \in H^{(2)}(H)\backslash H, \quad b \in \bar{H}^{(2)}(H)\backslash H\,. \tag{V 2}$$

Wenn nämlich immer $a \in H^{(1)}(H)\backslash H \Rightarrow a \notin \bar{H}^{(1)}(H)\backslash H$ und gleichzeitig $b \in H^{(2)}(H)\backslash H \Rightarrow$ $\Rightarrow b \notin H^{(2)}(H)\backslash H$, bedeutet das, daß sich die zwei Mengenpaare $(H^{(1)}(H); H^{(2)}(H))$ und $(\bar{H}^{(1)}(H); \bar{H}^{(2)}(H))$ nur in der Reihenfolge ihrer Komponenten unterscheiden.

Nehmen wir an, daß die Punkte a, b die Eigenschaften (V1), (V2) besitzen. Wegen (E2) folgt aus (V1): $a \in \bar{H}^{(2)}(H)\backslash H$. Wir haben also $a \in \bar{H}^{(2)}(H)\backslash H$, $b \in \bar{H}^{(2)}(H)\backslash H$, und deshalb muß $(ab) \cap H = \emptyset$ sein (sonst erhalten wir einen Widerspruch zu (E5) für die Menge $\bar{H}^{(2)}(H)$). Es ist aber gleichzeitig $a \in H^{(1)}(H)\backslash H$, $b \in H^{(2)}(H)\backslash H$, und die Menge $(ab) \cap H$ ist also wegen (E4) nichtleer. Aus diesem Widerspruch ergibt sich, daß Elemente a, b mit den Eigenschaften (V1), (V2) nicht existieren können und daß sich die angeführten Mengenpaare nur in der Reihenfolge unterscheiden können. ∎

Literatur

[1] Bryant, V., and R. Webster: Convexity spaces I: The basic properties, J. Math. Anal. Appl. **37** (1972) 1.

[2] Prenowitz, W., and J. Jantosciak: Geometries and join spaces, J. für Reine und Angew. Math. **257** (1972).

[3] Heidekrüger, G.: Untersuchungen in verallgemeinerten Konvexitätsräumen, Dissertation, Pädagogische Hochschule Halle „N. K. Krupskaja", 1976.

[4] Zimmermann, K.: Extremální algebra a zobecnění konvexnosti, Forschungsarbeit MFF UK, 1978.

Summary

An attempt to develope an axiomatics for a generalized convexity concept is made. The concept of a hyperplane is introduced axiomatically. A new concept of a generalized linearity is considered and some properties of optimization problems that are linear in this sense, are studied. These properties were made use of in literature in the process of solving certain class of optimization problems. The properties give us also a methodology for the construction of algorithms for the solving of some further optimization problems of this kind.

Резюме

Сделана попытка ввести аксиоматическое понятие обобщённой выпуклости и понятие гиперплоскости. Дальше вводится понятие некоторой обобщенной линейности и исследуются свойства оптимизационных задач, которые являются линейными по отношению к этому обобщённому понятию линейности. Эти свойства использовались в литературе при численном решении определенных классов оптимизационных задач. Они дают нам тоже методологию для построения алгоритмов для решения дальнейших оптимизационных задач этого рода.

Eingereicht bei der Redaktion: 7. 12. 1977; revidierte Fassung: 10. 5. 1978

Book Review

B. ANDRÁSFAI: **Introductory Graph Theory.** Akadémiai Kiadó, Budapest 1977, 266 S., Ft 170,–

Es wird eine Übersetzung eines 1969 im ungarischen Akademieverlag erschienenen Buches vorgelegt, des ersten Lehrbuches über Graphentheorie in ungarischer Sprache.
Dieser Teil ist vornehmlich der Theorie gewidmet, während ein in Vorbereitung befindlicher zweiter Teil sich auch den Anwendungen widmen wird.
Eine Fülle von Übungsaufgaben und Problemen bilden den eigentlichen Kern der einzelnen Kapitel. Der Leser wird, will er sich in den gebotenen Stoff einarbeiten, gezwungen, ständig Aufgaben zu lösen. Bekannte Sätze der Graphentheorie werden dem Leser als Probleme gestellt, die dann im weiteren Verlauf des Kapitels oder auch erst in einem Schlußkapitel teilweise oder ganz gelöst werden.
Neben einem Einführungskapitel werden in weiteren Kapiteln behandelt: Wälder, EULER-linien, HAMILTONlinien, Faktorisierung, Extremalaufgaben. Das Kapitel VI (Extremal-values, extremalgraphs) ist nicht nur das umfangreichste, sondern bildet den Hauptteil dieses Buches. Beim Lesen erkennt man unschwer die ungarische Graphenschule, die von D. KÖNIG in den 30er Jahren begründet wurde.
Kapitel VII ist ausschließlich der Lösung derjenigen Aufgaben und Probleme gewidmet, die im Textverlauf der einzelnen Kapitel noch nicht gelöst wurden.
Da im Text selber die Namen der Autoren der entsprechenden Sätze nicht angegeben sind, wird dieses in einem Quellenindex nachgeholt. Ein nicht allzu umfangreiches Literaturverzeichnis schließt dieses originell gestaltete Lehrbuch ab.
Gewidmet ist dieses Buch der verstorbenen ungarischen Mathematikerin PÉTER RÒZSA, die dem deutschen Leser durch ihr deutsches Büchlein „Spiel mit dem Unendlichen“, Teubner 1955, wohlbekannt ist.

H. WALTHER

Math. Operationsforsch. Statist., Ser. Optimization, Vol. 10 (1979) No. 1, 27–37

Basishäufigkeit bei linearen Restriktionen[1]

J. HIRCHE[2]

Zusammenfassung: Für den Fall, daß die Elemente einer $(m \times n)$-Matrix mit $m < n$ unabhängige gleichverteilte diskrete Zufallsgrößen mit nichtnegativen Realisierungen sind, werden Aussagen über die Wahrscheinlichkeit dafür gemacht, daß eine beliebige Auswahl von m Spaltenvektoren eine Basis bildet. Außerdem wird die relative Basishäufigkeit bei Transportrestriktionen untersucht. Die erzielten Resultate geben Aufschluß über den Anwendungsbereich solcher Optimierungsverfahren für Probleme mit linearen Restriktionen, bei denen jede quadratische Untermatrix von Höchstordnung als regulär vorausgesetzt wird.

1. Einleitung

Bei der Behandlung von Optimierungsproblemen, deren Restriktionen lineare Gleichungen $\boldsymbol{Ax}=\boldsymbol{b}$ mit $[\boldsymbol{A}]=m\times n$ sind, nimmt man neben $m<n$ gewöhnlich $\mathrm{rg}(\boldsymbol{A})=m$ an. Statt dessen wird in verschiedenen Arbeiten über Optimierungsverfahren vorausgesetzt, daß jede quadratische Untermatrix von Höchstordnung regulär ist (CANDLER und TOWNSLEY [1], KÜNZI und KLEIBOHM [8], RANI und SHIVPURI [11]). Eine rechteckige Matrix mit der letztgenannten Eigenschaft soll als *vollregulär* bezeichnet werden. Aus approximationstheoretischer Sicht bedeutet dies das Erfülltsein der HAARschen Bedingung (CHENEY [2], S. 45).

Die Beantwortung der Frage, in welchem Maße Vollregularität in Abhängigkeit von der Beschaffenheit der Koeffizientenmatrix erwartet werden kann, ist bedeutsam für die Einschätzung darauf beruhender Verfahren.

Bei der Untersuchung allgemeiner linearer Restriktionen werden die Koeffizienten hier im Hinblick auf die entsprechende Erzeugung von Testbeispielen als unabhängige gleichverteilte diskrete Zufallsgrößen vorausgesetzt. Unter Beschränkung auf derartige Zufallsgrößen mit nichtnegativen Realisierungen werden Aussagen über die Wahrscheinlichkeit dafür gemacht, daß eine Auswahl von m Spaltenvektoren eine Basis bildet. Bei den Nebenbedingungen des Transportproblems ist indessen das Verhältnis zwischen der Anzahl der Basislösungen und

---

[1] Vortrag, gehalten auf der Tagung „Mathematische Optimierung – Theorie und Anwendungen" in Eisenach (DDR), 5.–10. Dezember 1977, gekürzte Fassung von [4].

[2] Sektion Mathematik der Martin-Luther-Universität, DDR – 401 Halle (Saale), Universitätsplatz 6.

der aller möglichen Kombinationen von $m+n-1$ Koeffizientenvektoren zu betrachten, wobei m und n jetzt wie üblich die Anzahl der Versand- bzw. Empfangsorte bedeuten.

GIRKO [3], S. 11, betont bezüglich quadratischer Matrizen, deren Elemente unabhängige Zufallsgrößen sind, daß die Verteilungsfunktionen der Determinanten solcher Matrizen kompliziert sind und sich das Interesse deshalb auf die Untersuchung von Momenten konzentriert.

Allerdings konnte KOMLÓS [6] zeigen, daß die Wahrscheinlichkeit für das Verschwinden einer Determinante aus unabhängigen gleichverteilten Zufallsgrößen mit den Realisierungen 0 und 1 bei wachsender Ordnung der Determinante gegen Null strebt. Später hat er diese Aussage auf unabhängige Zufallsgrößen mit einer gemeinsamen nichtausgearteten Verteilungsfunktion verallgemeinert ([7]). Sein Vorgehen liefert aber keinen quantitativen Aufschluß über die Größe der genannten Wahrscheinlichkeit in Abhängigkeit von der Verteilungsfunktion und der Ordnung der Determinante. Dazu sollen die folgenden Überlegungen beitragen.

2. Aussagen zum allgemeinen Fall

Es werden m-dimensionale Vektoren betrachtet, deren Komponenten unabhängige gleichverteilte Zufallsgrößen mit der Realisierungsmenge $\{0, 1, \ldots, w-1\}$ sind. Lineare Abhängigkeit von m Vektoren liegt genau dann vor, wenn die aus ihnen gebildete Determinante verschwindet.

Im Falle $m=2$ gilt $\det((a_{il}))=a_{11}a_{22}-a_{12}a_{21}$, und die auftretenden Produkte sind unabhängige Zufallsgrößen. Tritt ein Produktwert $j\in\{0, 1, \ldots, (w-1)^2\}$ insgesamt x_j-mal auf, so ergibt sich mit $p_{jw}=x_j/w^2$ für die Wahrscheinlichkeit Q_w von $\det((a_{il}))=0$

$$Q_w=\sum_{j=0}^{(w-1)^2} p_{jw}^2 .$$

Dafür gelingt folgendermaßen eine Abschätzung: Die Tafel aller möglichen Produkte zweier a_{il} enthält w^2 Elemente, davon $2w-1$ Nullen. Von den von Null verschiedenen Elementen kommt keines in einer Reihe zweimal vor. Damit liegt der in Q_w eingehende Wert $f_w^{(0)}$ von

$$f(\boldsymbol{x})=\sum_{j=1}^{(w-1)^2} x_j^2 \tag{1}$$

zwischen dem Minimum und dem Maximum dieser konvexen Funktion unter den Nebenbedingungen

$$\sum_{j=1}^{(w-1)^2} x_j=(w-1)^2, \tag{2}$$

$$x_j\leqq w-1, \quad j=1(1)\,(w-1)^2, \tag{3}$$

$$x_j\geqq 0 . \tag{4}$$

Der dem Koordinatenursprung nächstgelegene Punkt (1, 1, ..., 1) der Hyperebene (2) erfüllt auch (3) und (4) und liefert

$$(w-1)^2 \leqq f_w^{(0)}. \qquad (5)$$

Andererseits nimmt $f(\boldsymbol{x})$ sein Maximum über dem zulässigen Bereich (2) bis (4) in einem Eckpunkt an. Die Ungleichungen (3) werden durch Einführung von Schlupfvariablen $y_j \geqq 0$ in Gleichungen überführt:

$$x_j + y_j = w-1, \quad j = 1(1)\,(w-1)^2. \qquad (6)$$

Die Koeffizientenmatrix von (2) und (6) hat die Gestalt

$$\tilde{\boldsymbol{A}} = \left[\begin{array}{c|c} 1 \ldots 1 & 0 \ldots 0 \\ \boldsymbol{E} & \boldsymbol{E} \end{array}\right],$$

wobei $\boldsymbol{E}$ die Einheitsmatrix der Ordnung $(w-1)^2$ ist. Die letzten $r = (w-1)^2 + 1$ Spalten von $\tilde{\boldsymbol{A}}$ lassen erkennen, daß $\tilde{\boldsymbol{A}}$ den Rang r hat.

Wegen (2) und (3) müssen in einer zulässigen Basislösung mindestens $w-1$ Variable x_j positiv sein. Wären $w+s$ Variable x_j mit $s \geqq 0$ positiv, so müßte $0 < x_j < w-1$ und $y_j > 0$ für mindestens zwei Indizes j gelten, und es läge keine Basislösung vor. Daher enthält jede zulässige Basislösung genau $w-1$ positive x_j mit $x_j = w-1$. Für jede wird der gesuchte Maximalwert $f = (w-1)^3$ angenommen, und mit (5) gilt

$$(w-1)^2 \leqq f_w^{(0)} \leqq (w-1)^3.$$

Unter Berücksichtigung der $2w-1$ Nullen in der Produkttafel folgt daraus die Abschätzung

$$\underline{\mathsf{Q}}_w = (5w^2 - 6w + 2)/w^4 \leqq \mathsf{Q}_w \leqq (w^3 + w^2 - w)/w^4 = \bar{\mathsf{Q}}_w. \qquad (7)$$

Also ist die Folge $\{\mathsf{Q}_w\}$ eine Nullfolge, und die Schranken $\underline{\mathsf{Q}}_w$ und $\bar{\mathsf{Q}}_w$ geben Aufschluß über die Konvergenzgeschwindigkeit. Zugleich sind $\bar{\mathsf{P}}_w = (1 - \underline{\mathsf{Q}}_w)$ und $\underline{\mathsf{P}}_w = (1 - \bar{\mathsf{Q}}_w)$ eine obere bzw. eine untere Schranke für die Wahrscheinlichkeit P, mit der eine Basis erhalten wird. Praktisches Interesse besitzt der Fall $w = b^k$ mit natürlichen Zahlen $b \geqq 2$ und $k \geqq 1$, insbesondere mit $b = 2$ und $b = 10$. Es gilt

Satz 1: *Sind die Elemente $a_{il} \geqq 0$ unabhängige gleichverteilte b-adische Zufallszahlen mit k Stellen, so liegt die Wahrscheinlichkeit für das Nichterhalten einer Basis zwischen $2b^{-2k}$ und b^{-k+1}.*

Beweis: Für $b \geqq 2$ und $k \geqq 1$ gilt $b^{2k} - 2b^k \geqq 0$. Daraus folgt nach Multiplikation mit 3 und Addition von $2b^{2k}$ schließlich

$$(5b^{2k} - 6b^k + 2)/b^{4k} \geqq 2b^{-2k} = \underline{\underline{\mathsf{Q}}}.$$

Außerdem ergibt sich mittels elementarer Abschätzungen

$$(b^{3k} + b^{2k} - b^k)/b^{4k} \leqq b^{-k+1} = \bar{\bar{\mathsf{Q}}},$$

und (7) mit $w = b^k$ liefert die Behauptung. ▮

Mit $\mathsf{P} = 1 - \bar{\bar{\mathsf{Q}}}$, $\bar{\bar{\mathsf{P}}} = 1 - \underset{=}{\mathsf{Q}}$ und $w = 2^k$ erhält man aus (7) und Satz 1 Abschätzungen für die Werte von $\mathsf{P} = \mathsf{P}(k)$, insbesondere folgt $\mathsf{P}(1) = 3/8$.

Schwieriger ist die zweiseitige Abschätzung der Wahrscheinlichkeit S_m für den Erhalt einer Basis bei festem w in Abhängigkeit von der Dimension m. Sei etwa $w = 2$, so daß der von Komlós [6] qualitativ untersuchte Fall vorliegt. Für $m = 2$ gilt dann $\mathsf{S}_2 = 3/8$. Allgemein gibt es unter Berücksichtigung der Anordnung 2^{m^2} Kombinationen von m Vektoren, im Falle $m = 3$ also 512, davon mit 3 voneinander und vom Nullvektor verschiedenen Vektoren $7 \cdot 6 \cdot 5 = 210$. Es existieren genau 6 Ebenen, die je 3 Vektoren dieser Teilmenge – und damit je 6 Kombinationen mit Berücksichtigung der Anordnung – enthalten, so daß die Anzahl der Kombinationen mit linear unabhängigen Vektoren 174 beträgt. Daraus folgt $\mathsf{S}_3 = 174/512$, und $\{\mathsf{S}_m\}$ ist nicht durchgängig monoton wachsend. Für allgemeines m erhält man als obere Schranke für S_m

$$\bar{\mathsf{S}}_m = \prod_{i=1}^{m} [(2^m - i)/2^m] .$$

Es gilt für $m \geqq 1$

$$\bar{\mathsf{S}}_m < 1 \Big/ \left(\frac{2^m - m/2 + m/2}{2^m - m/2}\right)^m ,$$

und mittels der Bernoullischen Ungleichung ([9], Bd. 1, S. 130) ergibt sich

$$\bar{\mathsf{S}}_m < \frac{2^m - m/2}{2^m - m/2 + m^2/2} = \bar{\bar{\mathsf{S}}}_m ,$$

was keine allzu schnelle Annäherung von S_m an 1 erwarten läßt.

3. Untersuchung der Restriktionen des Transportproblems

Für die Anzahl der Basen gilt in Abhängigkeit von der Anzahl der Versand- bzw. Empfangsorte (vgl. [12], [14])

$$A(m, n) = m^{n-1} n^{m-1} .$$

Da der Rang der Koeffizientenmatrix gleich $m + n - 1$ ist, ergibt sich für die relative Häufigkeit der Basen

$$\mathsf{P}(m, n) = m^{n-1} n^{m-1} \Big/ \binom{mn}{m+n-1} . \tag{8}$$

Es sollen Aussagen über das Verhalten von $\mathsf{P}(m, n)$ bei wachsenden m und n gemacht werden. Aus Symmetriegründen genügt die Betrachtung von $m \leqq n$. Dabei ist der Fall $m = 1$ mit $\mathsf{P}(1, n) = 1$ uninteressant. Die $\mathsf{P}(m, n)$ als Elemente einer unendlichen Matrix auffassend, wird man vor allem nach dem Einfluß einer Vergrößerung von n bei festem m bzw. einer gleichzeitigen Vergrößerung von m und n sowie nach einer etwaigen Beziehung zwischen Elementen mit übereinstimmender Summe $m+n$ fragen. Auskunft darüber geben die folgenden Sätze 2–4.

Satz 2 (Zeilenmonotonie): *Für jedes $m \geqq 2$ ist die Folge $\{\mathsf{P}(m, n)\}$ eine für $n \geqq m$ streng monoton abnehmende Nullfolge.*

Beweis: Gemäß (8) gilt $\mathsf{P}(m, n) > 0$ und

$$\frac{\mathsf{P}(m, n+1)}{\mathsf{P}(m, n)} = \frac{m\,(n+1)^{m-1}\,(m+n)\,(mn-n-m+2)\ldots(mn-n-1)\,(mn-n)}{n^{m-1}\,(mn+1)\,(mn+2)\ldots(mn+m-1)\,(mn+m)}\,.$$

Darin sind $m-1$ Faktoren der Form

$$f_k = \frac{(n+1)\,[(m-1)\,(n-1)+k]}{n(mn+1+k)}\,, \quad k=1(1)m-1\,,$$

enthalten, die wie folgt abgeschätzt werden können: Aus der für $k=1(1)\,m-1$ gültigen Ungleichung $0 \leqq (m+n)\;(m-1-k)$ folgt durch Addition von $m^2n^2 - mn^2 - m^2 + m + kmn + km$ auf beiden Seiten über

$$m\,(mn^2 - n^2 - m + 1 + kn + k) \leqq (m-1)\,(mn^2 + n + kn)$$

zunächst

$$f_k \leqq (m-1)/m = 1/[1 + 1/(m-1)]\,.$$

Damit ergibt sich

$$\prod_{k=1}^{m-1} f_k \leqq \frac{4}{9} \quad \text{für} \quad m \geqq 3\,. \tag{9}$$

Ferner folgt aus der für $n \geqq m$ richtigen Ungleichung $m\,(n-m)+2>0$ über Addition von $m\,(m+n)$ auf beiden Seiten

$$m\,(m+n)/(mn+1) < 2\,. \tag{10}$$

Mit (9) und (10) gilt $\mathsf{P}\,(m, n+1)/\mathsf{P}(m, n) < 8/9$. Andererseits ergibt sich unmittelbar

$$\mathsf{P}\,(2, n+1)/\mathsf{P}(2, n) = (n+2)/(2n+1) \leqq 4/5 \quad \text{für} \quad n \geqq 2\,.$$

Also stellt $\{\mathsf{P}(m, n)\}$ für jedes feste $m \geqq 2$ eine Nullfolge dar, die für $n \geqq m$ streng monoton abnehmend ist.

Satz 3 (Diagonalverhalten): *Gilt $2n-4 \geqq m \geqq \frac{n}{2}+2$ mit $n \geqq 4$, so nimmt $\mathsf{P}(m, n)$ bei gleichmäßiger Zunahme von m und n streng monoton ab.*

Beweis: Aus Symmetriegründen genügt die Betrachtung von $n=m+a$ mit $a \geqq 0$, ganz. Dann ist

$$\frac{\mathsf{P}\,(m+1, n+1)}{\mathsf{P}(m, n)} = \frac{(m+1)^{m+a}\,(m+a+1)^m\,(2m+a)\,(2m+a+1)}{m^{m+a-1}\,(m+a)^{m-1}\,(m^2+am+1)\,(m^2+am+2)} \times \frac{(m^2+(a-2)\,m-a+2)\ldots(m^2+am)}{(m^2+am+3)\ldots(m^2+(a+2)\,m+a+1)}\,,$$

wobei der letzte Bruch je $2m+a-1$ Faktoren im Zähler und im Nenner hat.

Die für $0<r\leqq s$ und $t\geqq 0$ gültige Ungleichung

$$\frac{r}{s}\leqq\frac{r+t}{s+t} \tag{11}$$

liefert hier für $k=1(1)2m+a-1$ angesichts $k-1\geqq 0$

$$\frac{m^2+am+1-k}{m^2+am+2m+a+2-k}\leqq\frac{m\,(m+a)}{(m+1)\,(m+a+1)}.$$

Damit ergibt sich

$$\frac{\mathsf{P}\,(m+1,n+1)}{\mathsf{P}(m,n)}\leqq\frac{m^m\,(m+a)^{m+a}}{(m+1)^m\,(m+a+1)^{m+a}}\times\frac{(m+1)\,(m+a+1)\,(2m+a)\,(2m+a+1)}{(m^2+am+1)\,(m^2+am+2)}=\mathsf{S}(m,a)\,, \tag{12}$$

und es gilt

$$\mathsf{S}(m,a)\to 4/e^2 \quad \text{für jedes feste } a \text{ und } \quad m\to\infty\,. \tag{13}$$

Hinreichend für monotone Abnahme von $\mathsf{S}(m,a)$ bei festem a und $m\geqq 4$ ist

$$\frac{\mathrm{d}}{\mathrm{d}x}\left(\frac{(x+1)\,(x+a+1)\,(2x+a)\,(2x+a+1)}{(x^2+ax+1)\,(x^2+ax+2)}\right)<0 \quad \text{für} \quad x\geqq 4\,.$$

Da das Vorzeichen der Ableitung des Quotienten mit dem ihres Zählers übereinstimmt, genügt die Untersuchung des Polynoms

$$\begin{aligned}R(x,a)=&-10x^6+(8-30a-2a^2)\,x^5+(24+20a-41a^2-5a^3)\,x^4\\&+(32+48a-12a^2-32a^3-4a^4)\,x^3+(54+48a+16a^2-8a^3\\&-13a^4-a^5)\,x^2+(32+54a+8a^2-8a^3-4a^4-2a^5)\,x+4+16a\\&+11a^2-4a^3-3a^4\,,\end{aligned}$$

dessen Koeffizienten Polynome $p_k(a)$ sind. Diese besitzen nach der Cartesischen Zeichenregel ([13], S. 284) sämtlich höchstens eine positive Nullstelle und sind für alle größeren Argumente a negativ. Wie man leicht nachprüft, gilt

$$p_k(2)<0 \quad \text{für} \quad k=1(1)6 \quad \text{und} \quad p_0(2)=0\,.$$

Also fällt bei $a\geqq 2$ $R(x,a)<0$ für alle $x>0$ aus. Es bleiben die Fälle $a=0$ und $a=1$ mit

$$R(x,0)=-10x^6+8x^5+24x^4+32x^3+54x^2+32x+4\,,$$

$$R(x,1)=-10x^6-24x^5-2x^4+32x^3+96x^2+80x+24$$

zu betrachten. Diese Polynome in x sind wiederum rechts von ihrer einzigen positiven Nullstelle negativ. Es gilt

$$R(x,0)<0 \quad \text{für} \quad x\geqq 3 \quad \text{und} \quad R(x,1)<0 \quad \text{für} \quad x\geqq 2\,.$$

Daraus folgt: $\mathsf{S}(m,0)$ ist für $m\geqq 3$ streng monoton abnehmend. Wegen (13) fallen alle $\mathsf{S}(m,0)$ von einem bestimmten Index m_0 ab kleiner als 1 aus. Man findet

$$S(3, 0) > 1 > S(4, 0),$$

also

$$S(m, 0) < 1 \quad \text{für} \quad m \geqq 4. \tag{14}$$

Nun wird die Frage nach einem $m = m(a)$ gestellt, für das das monoton abnehmende $S(m, a)$ kleiner als 1 ausfällt, und gezeigt, daß dies auch im Falle $a \neq 0$ von $m = 4 + a$ geleistet wird. Für $a \geqq 1$ gilt

$$\frac{(5+a)(5+2a)}{(4+a)(4+2a)+2} = \frac{25+15a+2a^2}{18+12a+2a^2} \leqq \frac{40+16(a-1)+2a^2}{30+12(a-1)+2a^2} \leqq \frac{4}{3},$$

wobei die letzte Ungleichung dem Übergang auf die Kehrwerte in (11) entspricht. Andererseits ergibt sich unmittelbar

$$\frac{(8+3a)(9+3a)}{(4+a)(4+2a)+1} = \frac{72+51a+9a^2}{17+12a+2a^2} \leqq \frac{72+54a+9a^2}{16+12a+2a^2} = \frac{9}{2}.$$

Das Produkt der beiden zuletzt abgeschätzten Terme ist gemäß (12) durch

$$\left(\frac{5+a}{4+a}\right)^{4+a} \times \left(\frac{5+2a}{4+2a}\right)^{4+2a} = N$$

zu dividieren, und es gilt für $a \geqq 1$

$$N \geqq \left(\frac{6}{5}\right)^5 \times \left(\frac{7}{6}\right)^6 = \frac{117649}{18750} > 6.$$

Daraus folgt zusammen mit (14) $S(m, a) < 1$ für $m = 4 + a$, $a = 0, 1, \ldots$, d. h. $m \geqq n/2 + 2$, $n \geqq 4$, und (12) sowie die Symmetrie der $P(m, n)$ liefern die Behauptung von Satz 3. ▮

Aus Satz 2 und Satz 3 ergibt sich die

Folgerung 1: *Wird in einem Problem mit $4 \leqq n/2 + 2 \leqq m \leqq n$ m nicht stärker als n vergrößert, so nimmt $P(m, n)$ dabei ab.*

Satz 4 (Schrankensatz): *Für $k \geqq 3$ ist das Diagonalelement $P(k, k)$ eine obere Schranke für alle $P(m, n)$ mit $m + n = 2k$ und $m, n \geqq 2$.*

Beweis: Wegen der Symmetrie von $P(m, n)$ in m und n genügt es, die Behauptung für $m = k - s$, $n = k + s$ und $s = 1(1)\, k - 2$ nachzuweisen. Zu zeigen ist also

$$\frac{P(k-s, k+s)}{P(k, k)} = \frac{(k-s)^{k+s-1}\,(k+s)^{k-s-1}\,((k-1)^2 - s^2)!\,k^2!}{k^{2k-2}\,(k^2 - s^2)!\,(k-1)^2!} \leqq 1 \tag{15}$$

$$\text{für} \quad k \geqq 3 \quad \text{und} \quad s = 1(1)\, k - 2.$$

Dazu werden die Fakultäten in (15) gemäß der STIRLINGschen Formel ([9], Bd. 3, S. 150) ohne Benutzung von Zusatzinformationen über ϑ_n ersetzt. Für den Quotienten in (15) erhält man mit $\vartheta_n = 1$ im Zähler und $\vartheta_n = 0$ im Nenner nach Um-ordnen die Ungleichung

$$\mathsf{Q} \leqq \left(\frac{k}{k-1}\right)^{2(k-1)^2+1} \left[\left(\frac{k-1-s}{k-s}\right)\left(\frac{k-1+s}{k+s}\right)\right]^{(k-1)^2-s^2+0,5} \tag{16}$$
$$\times \left(\frac{k}{k+s}\right)^{k+s} \left(\frac{k}{k-s}\right)^{k-s} \exp\{1/4\,((k-1)^2-s^2)+1/4k^2\}.$$

Durch Logarithmieren gehen die in (16) auftretenden Quotienten in Differenzen von Logarithmen über, die durch Integrale dargestellt und durch rationale Ausdrücke abgeschätzt werden können:

$$\ln \mathsf{Q} \leqq [2\,(k-1)^2+1] \ln \frac{k}{k-1} + [(k-1)^2-s^2+0,5] \tag{17}$$
$$\times \ln \left(\frac{k-1-s}{k-s}\right)\left(\frac{k-1+s}{k+s}\right) + (k+s) \ln \frac{k}{k+s} + (k-s) \ln \frac{k}{k-s}$$
$$+ \frac{1}{4\,((k-1)^2-s^2)} + \frac{1}{4k^2}.$$

Als Hilfsmittel dient die Keplersche Regel ([10], S. 197f.) mit $f(x)=1/x$. Unter Benutzung der für $k \geqq 12$ durchweg gültigen Ungleichungen (Beweise in [4])

$$\frac{1}{4\,((k-1)^2-s^2)} + \frac{1}{4k^2} \leqq \frac{1}{8k-12} + \frac{1}{4k^2} \leqq \frac{1}{6k}, \quad \frac{1}{6\,(k-1)} \leqq \frac{1}{5k}, \quad -\frac{8ks^2}{12k^2-3s^2} \leqq -\frac{2s^2}{3k},$$
$$\frac{s^6}{60k^5} \leqq \frac{s^2}{60k}, \quad \frac{1}{5k} + \frac{6k-1}{4k^2-2k} + \frac{1}{6k} \leqq \frac{2}{k}, \quad \frac{2k^2-1,5k}{3k^2-3s^2} \leqq \frac{10k+3s^2}{15k},$$
$$\frac{4k^2-7k+2,5}{3k^2-3k+0,75-3s^2} \leqq \frac{8k+3s^2}{6k}$$

ergibt sich dabei aus (17)

$$\ln \mathsf{Q} \leqq \frac{2}{k} - \frac{17s^2}{60k} + \frac{-20\,(k-1)\,(k-1-s)^2+k-1+s}{120\,(k-1+s)\,(k-1-s)^3} + \frac{-4s\,(k-1-s)+1}{240\,(k-1-s)^5}.$$

Wegen $1 \leqq s \leqq k-2$ und $(k-1-s)^2 \geqq 1$ sowie $k-1+s < 2\,(k-1)$ sind die Zähler der beiden letzten Brüche und damit diese selbst nicht positiv, so daß für $s \geqq 3$ bei $k \geqq 12$ die Beziehung $\mathsf{Q}<1$ folgt. Für $s=1$ und $s=2$ läßt sich (15) für $k \geqq k_0(s)$ direkt herleiten:

a) $$\frac{\mathsf{P}\,(k-1,\,k+1)}{\mathsf{P}(k,\,k)} = \left(\frac{(k-1)\,(k+1)}{k^2}\right)^{k-2} < 1 \quad \text{für} \quad k \geqq 3.$$

b) $$\frac{\mathsf{P}(k-2,\,k+2)}{\mathsf{P}(k,\,k)} = \frac{(k-2)^{k+1}\,(k+2)^{k-3}\,(k^2-3)\,(k^2-2)\,(k^2-1)\,k^2}{k^{2k-2}\,(k^2-2k-2)\,(k^2-2k-1)\,(k^2-2k)\,(k^2-2k+1)}.$$

Es gilt für $d=0(1)\,3$

$$(k-2)\,(k^2-d) = k^3-2k^2-dk+2d \leqq k\,(k^2-2k+1-d) = k^3-2k^2+k-dk$$

genau dann, wenn $2d \leqq k$ ausfällt. Daraus folgt für $k \geqq 6$

$$\frac{\mathsf{P}\,(k-2,\,k+2)}{\mathsf{P}(k,\,k)} \leqq \left(\frac{(k-2)\,(k+2)}{k^2}\right)^{k-3} < 1\,.$$

Schließlich ergibt sich die Gültigkeit von $\mathsf{Q} \leqq 1$ für $5 \leqq k \leqq 11$ und $3 \leqq s \leqq k-2$

sowie für $k=4$ und $k=5$ bei $s=2$ durch direkten Vergleich entsprechender $\mathsf{P}(m, n)$, womit (15) vollständig bewiesen ist. ∎

Unabhängig von der Durchführbarkeit einer analogen Abschätzung für $\mathsf{P}(m, n)$ mit ungerader Summe $m+n$ liefern obiger Schrankensatz, die Reihenmonotonie von $\mathsf{P}(m, n)$ sowie $\mathsf{P}(2, 3) < \mathsf{P}(2, 2)$ die

Folgerung 2: *Für* $m+n=2k+1 \geqq 5$ *mit* $m, n \geqq 2$ *gilt* $\mathsf{P}(m, n) < \mathsf{P}(k, k)$.

Im übrigen treffen die auf die Sachlage bei kleinen Werten von m und n gestützten Vermutungen, daß die $\mathsf{P}(m, n)$ für jedes feste m ab $n=2$ sowie die $\mathsf{P}(k-s, k+s)$ bei festem k mit wachsendem $s=0(1)\,k-2$ monoton abnehmend sind, nicht zu. Es gilt beispielsweise $\mathsf{P}(3, 19) < \mathsf{P}(2, 20)$. Daß ein solcher Randeffekt nicht auf die zweite Zeile beschränkt ist, kann leicht gezeigt werden, etwa mittels

$$\lim_{k \to \infty} \frac{\mathsf{P}(3, 2k-3)}{\mathsf{P}(4, 2k-4)} = \infty\,.$$

Die $\mathsf{P}(m, n)$ wurden für $2 \leqq m \leqq n \leqq 20$ berechnet. Ihr Verhalten veranschaulicht Abb. 1; die dort eingezeichneten Kurven grenzen Werte, die größer als die angegebenen Zehnerpotenzen sind, von kleineren ab.

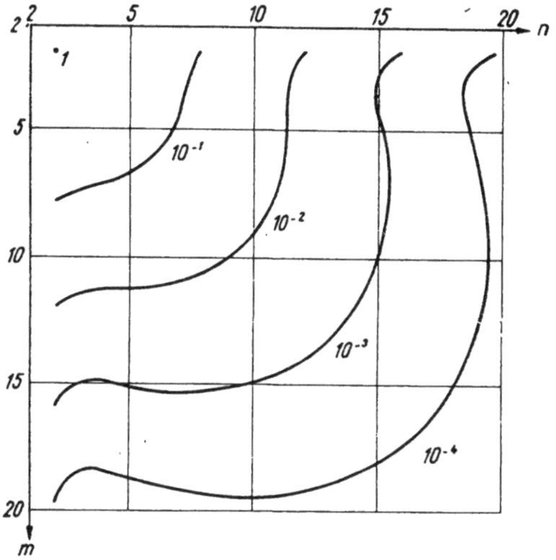

Abb. 1. Abnahme von $\boldsymbol{P}(m, n)$ in Abhängigkeit von m und n. (Der Achsenschnittpunkt in der Abbildung entspricht dem Punkt $m=n=1$ und nicht $m=n=2$.)

4. Konsequenzen für die Bestimmung einer guten Anfangsbasislösung

Mit dem Triplexverfahren wurde von Künzi und Kleibohm [8] eine Methode zur Lösung linearer Optimierungsprobleme vorgeschlagen, bei der das Ziel einer ersten Phase in der Bestimmung einer (nicht notwendig zulässigen) Basislösung in der Nähe des gesuchten Optimums besteht. Die Restriktionen einschließlich etwaiger Vorzeichenbeschränkungen für die Variablen werden als Nichtnegativitätsforderungen an Schlupfvariable in der Form

$$\boldsymbol{y}=\boldsymbol{b}+\boldsymbol{A}(-\boldsymbol{x})\geqq\boldsymbol{0} \tag{18}$$

angesetzt. Das Verfahren gründet sich auf die Annahme, daß man eine gute Anfangsbasislösung erhält, wenn diejenigen n Restriktionen aus (18) als Gleichungen erfüllt sind, deren Normalenvektoren mit dem Gradienten der Zielfunktion die kleinsten Winkel einschließen. Dabei wird stillschweigend vorausgesetzt, daß nicht nur die Matrix $\boldsymbol{A}$, sondern auch die Koeffizientenmatrix des so bestimmten Gleichungssystems Höchstrang hat und demzufolge tatsächlich eine Triplex*basis* resultiert. Andernfalls versagt das Triplexverfahren.

Nimmt man an, daß die Werte der Zielfunktionskoeffizienten unabhängig von Beziehungen zwischen den Elementen von $\boldsymbol{A}$ sind und letztere als Realisierungen diskreter gleichverteilter Zufallsgrößen aufgefaßt werden können, so ist ein solches Versagen nach den Ausführungen im Abschnitt 2 mit größerer Wahrscheinlichkeit bei Problemen von niedriger Dimension und mit kleiner Stellenzahl der Koeffizienten zu erwarten.

Dagegen ist ein Versagen des Triplexverfahrens bei den mit Hilfe von Zufallszahlen erzeugten Testbeispielen höherer Dimension, die in [8] ausgewertet wurden, offenbar nicht eingetreten.

Bei Problemen spezieller Struktur muß jedoch in besonderem Maße damit gerechnet werden. Das wurde von Wurzel [15] an Beispielen für das Transportproblem und das dazu duale Problem bestätigt und findet für den letzten Fall seine Begründung in den Ergebnissen von Abschnitt 3.

Im Interesse der Funktionssicherheit anzuwendender Lösungsverfahren sollte daher bei der Behandlung von Optimierungsproblemen mit streng quasimonotonen Zielfunktionen und linearen Restriktionen mittels eines Simplexalgorithmus auf bewährte Startverfahren der linearen Optimierung zurückgegriffen werden. Bei spezieller Struktur der Nebenbedingungen lassen zielfunktionsorientierte Ausschöpfungsverfahren gute Anfangslösungen erwarten. Für hyperbolische Transportprobleme wurde ein solches Verfahren in [5] dargestellt.

Literatur

[1] Candler, W., and R. J. Townsley: The Maximization of a Quadratic Function of Variables Subject to Linear Inequalities. Man. Sci. **10** (1964), 515–523.

[2] Cheney, E. W.: Introduction to Approximation Theory. McGraw-Hill Book Company, New York-St. Louis-San Francisco-Toronto-London-Sydney 1966.

[3] Гирко, В. Л.: Случайные матрицы. Изд. „Вища школа“, Киев 1975.

[4] Hirche, J.: Anhang zu: Nichtkonvexe Optimierungsprobleme mit zusammengesetzten Zielfunktionen. Diss. B, Martin-Luther-Universität Halle 1977.

[5] Hirche, J., J. Köhler und V. Stiebitz: Ein effektives Lösungsverfahren für das hyperbolische Transportproblem. Beiträge z. Anal. **7** (1975), 151–156.

[6] Komlós, J.: On the Determinant of (0, 1) Matrices. Studia Sci. Math. Hung. **2** (1967), 7–21.

[7] Komlós, J.: On the Determinant of Random Matrices. Studia Sci. Math. Hung. **3** (1968), 387–399.

[8] KÜNZI, H. P., und K. KLEIBOHM: Das Triplex-Verfahren. Unternehmensforschung **12** (1968), 145–154.
[9] v. MANGOLDT, H., und K. KNOPP: Einführung in die höhere Mathematik. S. Hirzel Verlag, 9. Aufl., Leipzig 1954.
[10] POLOSHI, G. N. (Hrsg.): Mathematisches Praktikum. B. G. Teubner Verlagsgesellschaft, Leipzig 1963.
[11] RANI, O., and S. SHIVPURI: An Algorithm for Linear Fractional Functionals Programming Problems. ZAMM **57** (1977), 75–80.
[12] SIMONNARD, M. A., and G. F. HADLEY: Maximum Number of Iterations in the Transportation Problem. Naval Res. Logist. Quart. **6** (1959), 125–129.
[13] WILLERS, F. A.: Methoden der praktischen Analysis. W. de Gruyter & Co., 2. Aufl., Berlin 1950.
[14] WINTGEN, G.: Über die Anzahl der möglichen nichtausgearteten Basislösungen beim Transportproblem. Wiss. Z. Hochsch. Arch. u. Bauw. Weimar **11** (1964) 1, 78–79.
[15] WURZEL, B.: Anwendung von Duoplex- und Triplex-Verfahren auf lineare Optimierungsprobleme spezieller Struktur und Effektivitätsvergleich mit anderen Methoden. Diplomarbeit an der Sektion Mathematik der Martin-Luther-Universität, Halle 1973.

Summary

For the case that the elements of an $m \times n$ matrix with $m < n$ are independent discrete uniformly distributed random variables with nonnegative realizations statements on the probability that some choice of m column vectors forms a basis are given. Moreover, the relative frequency of bases in the constraints of transportation problems is studied. Results obtained permit conclusions on the applicability of such programming methods for problems with linear constraints where every quadratic submatrix of highest order is assumed to be regular.

Резюме

Предполагая, что элементы $m \times n$-матрицы $(m < n)$ являются дискретными, независимыми, равномерно распределенными случайными величинами с неотрицательными реализациями, получены результаты о вероятности того, что произвольная выборка m столбцов образует базис. Кроме того исследуется относительная частота базисов в транспортных задачах. Полученные результаты позволяют сделать выводы об области применения таких методов оптимизации для задач с линейными условиями, у которых каждая квадратичная подматрица, имеющая максимальный порядок, предполагается регулярной.

Eingereicht bei der Redaktion: 26. 1. 1978; revidierte Fassung: 25. 4. 1978

Book Review

A. Auslender (ed.): **Convex Analysis and Its Applications.** Proceedings of a Conference Held at Murat-le-Quaire, March 1976. Lecture Notes in Economics and Mathematical Systems 144. Springer-Verlag, Berlin-Heidelberg-New York 1977, 219 S., DM 24,80

In diesen Band wurden Vorträge aufgenommen, die anläßlich der Tagung über konvexe Analysis in Murat-le-Quaire (1.–6. März 1976) gehalten wurden. Diese Tagung sollte den gegenwärtigen Stand der Forschung an den französischen Universitäten auf dem Gebiet der konvexen Analysis und ihrer Anwendungen demonstrieren.

Die in diesem Band vereinigten Arbeiten überstreichen ein großes Spektrum. Neben Arbeiten, die sich mit algebraischen bzw. topologischen Grundlagen der konvexen Analysis beschäftigen, wurden auch solche aufgenommen, die sich mit Anwendungen in der nichtlinearen Optimierung, der Variationsrechnung und der Theorie der optimalen Steuerung befassen. Hier reicht die Breite von Stabilitätsuntersuchungen über Optimalitätsbedingungen, Dualität bis hin zu Verfahren. Im einzelnen findet man die folgenden Beiträge: M. Atteia und J. Audounet: Noyaux des sous-espaces de Banach reflexif d'un espace localement convexe; B. Cornet: An abstract theorem for planning procedures; P. Charrier, B. Hanouzet und J. L. Joly: Estimation and regularity for the maximum lower solution of a unilateral problem; J. P. Crouzeix: Conjugacy in quasiconvex analysis; O. Debordes: Duality: some results in asymptotical elastoplasticity; R. Janin: Sensitivity for non convex optimization problems; P. J. Laurent und C. Carasso: A dual algorithm in quasi-convex optimization; J. Mossino: Etude d'une inequation quasi-variationelle apparaissant en physique; M. F. Nougues-Sainte-Beuve: Vector measures with bounded variation: a topological property and some applications; G. Owen: Computation of equilibrium price and quantity vectors; J. Saint-Pierre: Borel convex-Valued multifunctions; R. J. B. Wets: On the convergence of random convex sets; J. P. Zolesio: An optimal design procedure for optimal control support.

Der Inhalt der Arbeiten unterstreicht die wachsende Bedeutung und relative Eigenständigkeit, die der konvexen Analysis heute zuerkannt werden muß. Neben nützlichen Überblicken finden sich zahlreiche interessante neue Ergebnisse, die diesen Tagungsbericht für den in der Forschung Tätigen wertvoll machen.

K.-H. Elster

Math. Operationsforsch. Statist., Ser. Optimization, Vol. 10 (1979) No. 1, 39–55

Das Verhalten von Abstiegsverfahren an einer Singularität des Gradienten[1]

ANITA ILLGEN[2]

Zusammenfassung: Untersucht wird das Verhalten von Abstiegsverfahren zur Minimierung von Funktionalen, die im wesentlichen stark und beschränkt konvex sind außer in der Nähe der Singularität des Gradienten. Es werden in Abhängigkeit von Verfahrensparametern notwendige und hinreichende Konvergenzbedingungen aufgestellt und Aussagen über die Konvergenzgeschwindigkeit insbesondere in der Nähe der Singularität getroffen. Die Konvergenz des Standortalgorithmus kann i. allg. im Einzugsbereich des Minimums mit diesen Mitteln gezeigt werden.

1. Einleitung

Die Struktur des Standortproblems und das Verhalten des KUHNschen Algorithmus zur Lösung dieses Problems gaben uns die Anregung, das Verhalten von Abstiegsverfahren in der Nähe einer Singularität des Gradienten zu untersuchen. Ausgangspunkt waren dabei die Methoden, die bei Konvergenzuntersuchungen für Abstiegsverfahren zur Lösung freier Probleme mit stark und beschränkt konvexem Zielfunktional angewandt werden [1], [8]. Betrachtet man nun Zielfunktionale, die zwar im wesentlichen noch stark und beschränkt konvex sind, deren Gradient aber eine Singularität aufweist, so ist von Interesse, wie diese Singularität das Konvergenzverhalten der Abstiegsverfahren beeinflußt.

Als Grundtyp behandeln wir das freie, konvexe Optimierungsproblem

$$f(x)=\bar{f}(x)+p\,\|x-a\|=\underset{x\in D\subseteqq H}{\mathrm{Min}}\,!,\quad a\in D,\quad p>0 \tag{P}$$

über der offenen konvexen Teilmenge D des Hilbertraumes H mit $\partial D\cap\{x \mid f(x)\leqq \alpha\}=\emptyset$ für mindestens ein $\alpha\in R^1$. Das Funktional $\bar{f}$ sei dabei stetig sowie *stark und beschränkt konvex*, d. h. für $0<\bar{m}\leqq\bar{M}$ gilt

$$\frac{\bar{m}}{2}\lambda_1\lambda_2\|x^1-x^2\|^2\leqq\lambda_1\bar{f}(x^1)+\lambda_2\bar{f}(x^2)-\bar{f}(\lambda_1x^1+\lambda_2x^2) \tag{1.1}$$

$$\leqq\frac{\bar{M}}{2}\lambda_1\lambda_2\|x^1-x^2\|^2$$

für alle $x^1, x^2\in D$, $\lambda_1, \lambda_2\geqq 0$, $\lambda_1+\lambda_2=1$.

---

[1] Vortrag, gehalten auf der Tagung „Mathematische Optimierung – Theorie und Anwendungen" in Eisenach (DDR), 5.–10. Dezember 1977.

[2] Sektion Mathematik, Karl-Marx-Universität, 701 Leipzig – DDR, Karl-Marx-Platz.

Da $\bar{f}$ stark konvex und $p\,\|x-a\|$ konvex sind, ist f auch stark konvex mit $m=\bar{m}>0$. Aber f ist nicht mehr beschränkt konvex über D. Der Gradient von f besitzt in a eine Singularität.

Zur Lösung des Problems (P) betrachten wir Abstiegsverfahren. Die entsprechende Iterationsfolge ergibt sich rekursiv aus

$$x^{k+1}=x^k+t_k h^k\,, \tag{1.2}$$

wobei die t_k hinreichend kleine, positive Schrittweiten und die h^k brauchbare Richtungen sind.

Im glatten Fall ($p=0$) wird etwa für Verfahren mit gleichmäßig brauchbaren Richtungen und Optimalschrittweite lineare Konvergenz nachgewiesen (vgl. [1]). Zur Lösung des Problems (P) können sich solche Verfahren aber an einer nichtoptimalen singulären Stelle a „festfahren". Das wird ersichtlich am folgenden

Beispiel 1: Wir suchen das Minimum von $f(x)=\left\|x-\begin{pmatrix}2\\0\end{pmatrix}\right\|^2+\frac{7}{2}\|x\|$, $x\in R^2$. Das Abstiegsverfahren

$$x^{k+1}=x^k+t_k\begin{pmatrix}\tan\alpha\\(-1)^k C\end{pmatrix},\quad C>2+\tan^2\alpha,\quad \tan\alpha=\frac{1}{7}\sqrt{15}\,, \tag{1.3}$$

definiert einen Prozeß mit Optimalschrittweiten t_k und gleichmäßig brauchbaren Richtungen. Die Iterationsfolge beschreibt dann im Kegel $K^-=\{x \mid x\in R^2, |x_2|\leqq -x_1\tan\alpha,\ x_1\leqq 0\}$ eine Zickzackbewegung und konvergiert gegen die Stelle a (siehe Abb. 1).

Dieses Beispiel demonstriert, daß man in Anlehnung an den glatten Fall jetzt in der Nähe der Stelle a zusätzliche Forderungen an den Prozeß stellen muß, um seine Konvergenz zu garantieren. Das ist insbesondere dann wichtig, wenn der Optimalpunkt x^* nahe an der Stelle a liegt.

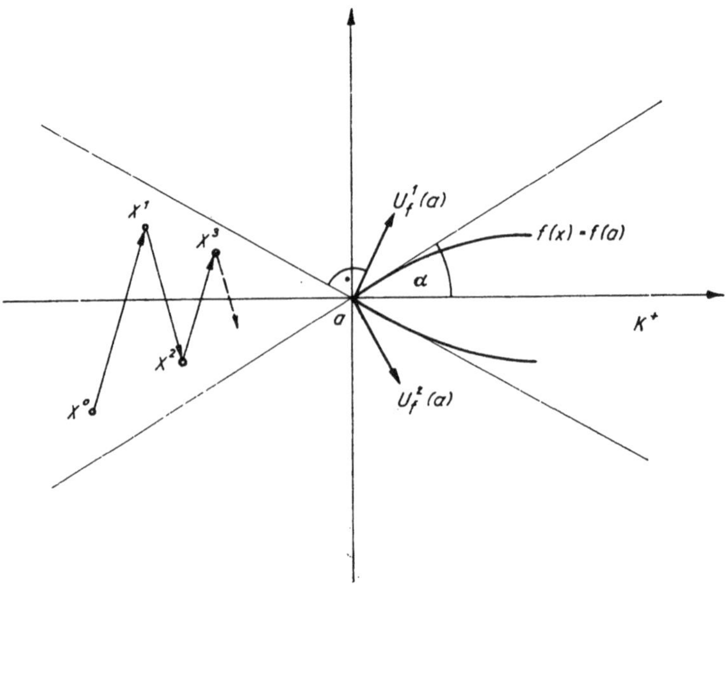

Abb. 1

2. Einschließungsaussagen für das Minimum

Wir stellen jetzt Mittel bereit, die für die Konvergenzuntersuchungen benötigt werden. Dazu gehören die Charakterisierung der Lage des Minimums in bezug auf die singuläre Stelle a und Vergleiche zwischen den Größen $\|x-x^*\|$, $f(x)-f(x^*)$ und $\|\nabla f(x)\|$, die die Abweichung zum Minimum messen.

Das Funktional f nimmt über D sein Minimum an. Das folgt daraus, daß f stark konvex und stetig und folglich alle Niveaumengen $N_f(\alpha)=\{x \mid f(x)\leqq\alpha\}$ beschränkt und abgeschlossen sind [2].

Eine notwendige und hinreichende Bedingung für den Minimalpunkt x^* konvexer, stetiger Funktionale ist die Existenz eines Subgradienten U_fx^* mit $U_fx^*=0$ (siehe [4]). Daraus ergibt sich: Für $x^*\neq a$ ist notwendig und hinreichend

$$\nabla\bar{f}(x^*)=p\frac{a-x^*}{\|a-x^*\|} \tag{2.1}$$

und für $x^*=a$ ist wegen

$$\begin{aligned}\langle\nabla\bar{f}(a), h\rangle+p\|h\| &= \lim_{t\to+0}\frac{f(a+th)-f(a)}{t} \\ &\geqq\langle U_fa, h\rangle=0\end{aligned} \tag{2.2}$$

mit $U_fa=0$ notwendig und hinreichend

$$\|\nabla\bar{f}(a)\|\leqq p\,. \tag{2.3}$$

Faßt man $p\|x-a\|$ als Störung der reduzierten Optimierungsaufgabe Min $\bar{f}(x)=$ $=\bar{f}(y^*)$ auf, dann wird durch p und die Lage von a die Lage des Optimalpunktes x^* von (P) beeinflußt. Ist $a=y^*$, so bleibt y^* optimal. Fällt a in $N_{\|\nabla\bar{f}\|}(p)$, so wird wegen (2.3) der Optimalpunkt nach a gezogen. Liegt a außerhalb von $N_{\|\nabla\bar{f}\|}(p)$, so „zieht" a den Optimalpunkt auf den Randpunkt von $N_{\|\nabla\bar{f}\|}(p)$, der auch Lotfußpunkt von a auf eine Niveaumenge von $\bar{f}$ ist.

In der Nähe von a beschreiben folgende Aussagen das Verhalten von $\|\nabla f(x)\|$. Und zwar ist

$$\begin{aligned}|\|\nabla\bar{f}(a)\|-p| &= \varliminf_{x\to a}\|\nabla f(x)\|=\lim_{t\to+0}\|\nabla f(a-t\nabla\bar{f}(a))\| \\ \|\nabla\bar{f}(a)\|+p &= \varlimsup_{x\to a}\|\nabla f(x)\|=\lim_{t\to+0}\|\nabla f(a+t\nabla\bar{f}(a))\|\,.\end{aligned} \tag{2.4}$$

Hieraus erhalten wir, daß für $\|\nabla\bar{f}(a)\|\neq p$ zu beliebig festem $x^0\in H$ ein $\varepsilon(x^0)>0$ existiert mit

$$\|\nabla f(x)\|\geqq\varepsilon>0 \tag{2.5}$$

für alle $x\in M_f(x^0)=\{x \mid f(a)\leqq f(x)\leqq f(x^0)\}$.

Wir benötigen im weiteren noch zu (1.1) äquivalente Bedingungen für die starke und beschränkte Konvexität eines Funktionals. Und zwar sind für stetige, konvexe Funktionale g gleichbedeutend (vgl. [9]):

a) g ist stark konvex über D. (2.6)

b) $\frac{m}{2} r^2 \leqq g(x+h) - g(x) - \langle U_g x, h\rangle \quad \forall x, x+h \in D, \quad \|h\| = r$,

c) $mr^2 \leqq \langle U_g(x+h) - U_g x, h\rangle \quad \forall x, x+h \in D, \quad \|h\| = r$.

Wegen (2.6) b) gilt dann für stark konvexe Funktionale stets

$$g(x) - g(x+h) \leqq \frac{\langle U_g x, h\rangle^2}{2m\,\|h\|^2}. \tag{2.7}$$

Dabei bezeichnet $U_g x$ einen beliebigen Subgradienten von g an der Stelle x. Für das Zielfunktional f von (P) sind mit

$$U_f a = \bar{f}(a) + \frac{\alpha s}{\|s\|}, \quad 0 \leqq \alpha \leqq p, \quad s \in H, \tag{2.8}$$

alle Subgradienten an der Stelle a gegeben.

Außerdem sind für stetige, konvexe Funktionale gleichbedeutend (vgl. [1]):

a) g ist beschränkt konvex über D. (2.9)

b) $g(x+h) - g(x) - \langle U_g x, h\rangle \leqq \frac{M}{2} r^2 \quad \forall x, x+h \in D, \quad \|h\| = r$,

c) $\langle U_g(x+h) - U_g x, h\rangle \leqq Mr^2 \quad \forall x, x+h \in D, \quad \|h\| = r$.

Jeder Subgradient U_g ist dann wegen (2.9) b) Fréchet-Gradient von g. Für $g = \|x-a\|$ ist z. B. mit

$$\|x + th - a\| \leqq \|x - a\| + t\left\langle \frac{x-a}{\|x-a\|}, h\right\rangle + \frac{1}{2\,\|x-a\|} t^2 \|h\|^2 \tag{2.10}$$

über jeder konvexen Menge, zu deren Abschluß a nicht gehört, (2.9) b) erfüllt.

Nun geben wir noch einige Abschätzungen an, die die verschiedenen Maße der Abweichung vom Minimum vergleichen. Für die stark konvexen Funktionale f gilt [9]:

$$\|U_f x^0\|^2 \geqq 2m\,(f(x^0) - f(x^*)) \geqq m^2 \|x^0 - x^*\|^2 \tag{2.11}$$

für jedes $x^0 \in D$ und jeden Subgradienten $U_f x^0$. Außerdem ist

$$\frac{1}{2} \|x^0 - a\| \frac{\|\nabla f(x^0)\|^2}{\bar{M}\,\|x^0 - a\| + p} \leqq f(x^0) - f(x^*) \tag{2.12}$$

für $x^0 \in D$ und $x^0 \neq a$. Diese Abschätzung erhalten wir aus der beschränkten Konvexität von $\bar{f}$ und (2.10) mit

$$f(x^0 - t\nabla f(x^0)) - f(x^0) \leqq \frac{1}{2} t^2 \|\nabla f(x^0)\|^2 \left(\bar{M} + \frac{p}{\|x^0 - a\|}\right) - t\|\nabla f(x^0)\|^2, \tag{2.13}$$

indem wir über beide Seiten das Minimum bilden.

3. Konvergenzuntersuchungen

Die Abstiegsverfahren (1.2) sind durch Vorgabe der Richtungen und Schrittweiten bestimmt. Um solche Vorschriften an diese Verfahrensparameter zu erhalten, die Konvergenz garantieren, untersuchen wir zunächst das Verhalten von f längs einer brauchbaren Richtung und das Verhalten von Abstiegsverfahren in der Nähe der Singularität a.

Es heißt eine *Richtung* h im Punkt x^0 *brauchbar*, wenn gilt:

$$\langle \nabla f(x^0), h\rangle < 0 \quad \text{für} \quad x^0 \neq a\,, \qquad (3.1)$$
$$\langle \nabla \bar{f}(a), h\rangle + p\,\|h\| < 0 \quad \text{für} \quad x^0 = a\,.$$

Der Parameter

$$c = \begin{cases} \dfrac{|\langle \nabla f(x^0), h\rangle|}{\|\nabla f(x^0)\| \cdot \|h\|} & \text{für} \quad x^0 \neq a \\ \dfrac{|\langle \nabla \bar{f}(a), h\rangle|}{\|\nabla \bar{f}(a)\| \cdot \|h\|} & \text{für} \quad x^0 = a \end{cases} \qquad (3.2)$$

beschreibt die *Güte* der brauchbaren Richtung h. Es heißen dann die Richtungen h^k *gleichmäßig brauchbar* in x^k, wenn $0 < c \leqq c_k$ für alle k gilt, und optimal brauchbar, wenn $c_k = 1$ für alle k gilt.

Längs brauchbarer Richtungen fallen die Funktionswerte. Durch die folgenden Schrittweitenvorschriften wird ein Mindestabfallen von f garantiert.

Satz 1: *Es sei h eine in x^0 brauchbare Richtung. Dann besitzt f längs des Strahls $x = x^0 + th$, $t \geqq 0$, ein eindeutiges Minimum $f(x^0 + t^*h) = \mathrm{Min}\,\{f(x^0 + th) \mid t \geqq 0\}$, und es gilt:*

$$f(x^0) - f\,(x^0 + th) \geqq \frac{\underline{M}}{2}\,\|h\|^2\,\underline{\sigma}^2 \quad \textit{für alle} \quad \underline{\sigma} \leqq t \leqq t^* \qquad (3.3)$$

bei

$$0 < \underline{\sigma} = \begin{cases} \dfrac{-1}{\bar{M}\,\|h\|^2}\left(\langle \nabla \bar{f}(x^0), h\rangle + p\left\langle \dfrac{x^0 + \underline{\sigma}h - a}{\|x^0 + \underline{\sigma}h - a\|}, h\right\rangle\right) \\ \dfrac{\|x^0 - a\|}{\|h\|} \quad \text{für} \quad x^0 + \underline{\sigma}h = a\,. \end{cases} \qquad (3.4)$$

Beweis: Wegen der starken Konvexität von f und der Brauchbarkeit der Richtung h wird das Infimum von f längs $x^0 + th$ eindeutig in $x^0 + t^*h$, $t^* > 0$, angenommen. Nun zeigen wir, $f(x^0) - f\,(x^0 + \underline{\sigma}h)$ ist durch $\bar{M}\,\frac{1}{2}\,\|h\|^2\,\underline{\sigma}^2$ nach unten beschränkt, und es gilt $0 < \underline{\sigma} \leqq t^*$. Daraus folgt dann, da f in $[0, t^*]$ monoton fällt, die Behauptung.

Wir unterscheiden drei Fälle:

Fall 1: $h \neq \lambda\,(a - x^0)$ für alle $\lambda \geqq 0$, $x^0 \neq a$. Aus der beschränkten Konvexität von $\bar{f}$ folgt

$$f\,(x^0 + th) - f(x^0) \leqq \frac{\bar{M}}{2}\,t^2\,\|h\|^2 + t\langle \nabla \bar{f}(x^0), h\rangle \qquad (3.5)$$
$$+ p\,\{\|x^0 + th - a\| - \|x^0 - a\|\} = v_0(t)\,.$$

Dabei nimmt $v_0(t)$ ihr Minimum für

$$\underline{\sigma} = \frac{-1}{\bar{M}\,\|h\|^2}\left(\langle \nabla \bar{f}(x^0), h\rangle + p\left\langle \frac{x^0 + \underline{\sigma}h - a}{\|x^0 + \underline{\sigma}h - a\|}, h\right\rangle\right) = s_0(\underline{\sigma}) \qquad (3.6)$$

an. Da $s_0(0) > 0$, $s_0(t)$ stetig und monoton streng fallend ist, existiert $\underline{\sigma} > 0$ eindeutig als Lösung von $s_0(t) = t$. Wir setzen noch die Konvexität von $\|x^0 + th - a\|$ ein:

$$\|x^0 - a\| - \|x^0 + \underline{\sigma} h - a\| \geqq -\underline{\sigma} \left\langle \frac{x^0 + \underline{\sigma} h - a}{\|x^0 + \underline{\sigma} h - a\|}, h \right\rangle . \tag{3.7}$$

Mit (3.5) erhalten wir

$$f(x^0 + \underline{\sigma} h) - f(x^0) \leqq v_0(\underline{\sigma}) \leqq -\frac{\bar{M}}{2} \underline{\sigma}^2 \|h\|^2 . \tag{3.8}$$

Um $\underline{\sigma} \leqq t^*$ zu zeigen, nutzen wir die Monotonie von $s_0(t)$ aus. Es ist

$$\underline{\sigma} \leqq t \leftrightarrow s_0(t) \leqq t . \tag{3.9}$$

Wegen

$$0 = \langle \nabla f(x^0 + t^* h), h \rangle = \langle \nabla \bar{f}(x^0 + t^* h), h \rangle + p \left\langle \frac{x^0 + t^* h - a}{\|x^0 + t^* h - a\|}, h \right\rangle \tag{3.10}$$

$$\langle \nabla \bar{f}(x^0 + t^* h) - \nabla \bar{f}(x^0), h \rangle \leqq \bar{M} t^* \|h\|^2$$

gilt $s_0(t^*) \leqq t^*$, also $\underline{\sigma} \leqq t^*$.

Fall 2: $h = \lambda (a - x^0)$ für ein $\lambda > 0$, $x^0 \neq a$. Dann nimmt (3.5) die Gestalt

$$f(x^0 + th) - f(x^0) \leqq \frac{\bar{M}}{2} t^2 \lambda^2 \|a - x^0\|^2 + t\lambda \langle \nabla \bar{f}(x^0), a - x^0 \rangle \tag{3.11}$$

$$+ p \{|1 - \lambda t| - 1\} \|a - x^0\|$$

an. Durch Unterscheiden der Fälle $t^* = \frac{1}{\lambda}$, $t^* \neq \frac{1}{\lambda}$ und bei letzterem wieder $\underline{\sigma} < \frac{1}{\lambda}$, $\underline{\sigma} = \frac{1}{\lambda}$ bzw. $\underline{\sigma} > \frac{1}{\lambda}$ erhalten wir mit gleichen Mitteln die Behauptung.

Fall 3: $x^0 = a$. Die beschränkte Konvexität von $\bar{f}$ liefert

$$v_0(t) = \frac{\bar{M}}{2} t^2 \|h\|^2 + t (\langle \nabla \bar{f}(a), h \rangle + p \|h\|) \tag{3.12}$$

$$\text{Min } v_0(t) = -\frac{(\langle \nabla \bar{f}(a), h \rangle + p \|h\|)^2}{2\bar{M} \|h\|^2} = -\frac{\bar{M}}{2} \underline{\sigma}^2 \|h\|^2$$

mit

$$\underline{\sigma} = -\frac{1}{\bar{M} \|h\|^2} (\langle \nabla \bar{f}(a), h \rangle + p \|h\|) . \tag{3.13}$$

Wegen

$$-p \|h\| = \langle \nabla \bar{f}(a + t^* h), h \rangle , \tag{3.14}$$

$$\langle \nabla \bar{f}(a + t^* h) - \nabla \bar{f}(a), h \rangle \leqq \bar{M} t^* \|h\|^2$$

ist wieder $\underline{\sigma} \leqq t^*$ gezeigt.

Die so bestimmte Schrittweite $\underline{\sigma}$ beschreibt schwache Unterrelaxation. Sie ist in $x^0 \neq a$ nur implizit gegeben, aber wegen der Monotonieeigenschaften von $s_0(t)$

kann man vorgegebene Schrittweiten auf Unterrelaxation testen, wie das für den folgenden Schrittweitentyp σ demonstriert wird.

Satz 2: *Es sei h eine in $x^0 \neq a$ brauchbare Richtung. Dann gilt*

$$f(x^0) - f(x^0 + th) \geqq \frac{1}{2} \left(\bar{M}(x^0) + \frac{p}{\|x^0 - a\|} \right) \sigma^2 \|h\|^2 \tag{3.15}$$

für alle $t \in [\operatorname{Min}\{\sigma, t^*\}, \operatorname{Max}\{\sigma, t^*\}]$ *bei gewissem* $\bar{m} \leqq \bar{M}(x^0) \leqq \bar{M}$ *und für alle* $t \in [\underline{\sigma}, t^*]$ bei $\bar{M}(x^0) = \bar{M}$.
Dabei ist

$$0 < \sigma = \begin{cases} \dfrac{-\langle \nabla f(x^0), h\rangle}{\left(\bar{M}(x^0) + \dfrac{p}{\|x^0 - a\|}\right) \|h\|^2} & \text{für } x^0 \neq a \\ \underline{\sigma} & \text{für } x^0 = a. \end{cases} \tag{3.16}$$

Beweis: Wegen (2.10) und der beschränkten Konvexität von $\bar{f}$ ist für alle $t \geqq 0$ und für ein $\bar{M}(x^0)$ mit $\bar{m} \leqq \bar{M}(x^0) \leqq \bar{M}$

$$f(x^0 + th) - f(x^0) \leqq \frac{1}{2} \left(\bar{M}(x^0) + \frac{p}{\|x^0 - a\|} \right) t^2 \|h\|^2 + t \langle \nabla f(x^0), h\rangle = w_0(t). \tag{3.17}$$

Dabei nimmt $w_0(t)$ ihr Minimum in σ aus (3.16) an mit

$$w_0(\sigma) = -\frac{1}{2} \left(\bar{M}(x^0) + \frac{p}{\|x^0 - a\|} \right) \sigma^2 \|h\|^2 . \tag{3.18}$$

Für $\bar{M}(x^0) = \bar{M}$ gilt außerdem $v_0(\underline{\sigma}) \leqq w_0(\sigma)$ mit $v_0(t)$ aus (3.5). Aus den Monotonieeigenschaften von f längs $x^0 + th$ folgt die Behauptung.

Die Schrittweite σ ist ein konstruktiver Verfahrensparameter. Mit der Wahl von $\bar{M}(x^0)$ gibt es in ihrer Darstellung noch einen Freiheitsgrad. Diese Schrittweite beschreibt gemischte Relaxation, da sie auch größer als die Optimalschrittweite sein kann. Dies zeigt das folgende

Beispiel 2: Sei $f(x) = \frac{1}{2} \|x - b\|^2 + \|x\|$, $x^0 = -b \neq 0$. Wir wählen als brauchbare Richtung $h = -\nabla f(x^0) = \frac{1}{\|b\|} (2\|b\| + 1)\, b$. Es gilt

$$\underline{\sigma} < \sigma \leftrightarrow s_0(\sigma) < \sigma \leftrightarrow \chi(\sigma) < \psi(\sigma) \tag{3.19}$$

bei

$$\varphi(\underline{\sigma}) = \bar{M} \underline{\sigma} \|h\|^2 + \langle \nabla \bar{f}(x^0), h\rangle = -\frac{p}{\|x^0 + \underline{\sigma} h - a\|} \langle x^0 + \underline{\sigma} h - a, h\rangle = \chi(\underline{\sigma}) \tag{3.20}$$

$$\varphi(\sigma) = \bar{M} \sigma \|h\|^2 + \langle \nabla \bar{f}(x^0), h\rangle = -\frac{p}{\|x^0 - a\|} \langle x^0 + \sigma h - a, h\rangle = \psi(\sigma)$$

für $a = 0$, $p = 1$, $\bar{M} = 1$. In unserem Beispiel ist $\underline{\sigma} = t^*$, und wegen $\langle x^0, h\rangle < 0$, $\langle x^0 + \sigma h, h\rangle > 0$ und $\|x^0 + \sigma h - a\| < \|x^0 - a\|$ ist $\chi(\sigma) < \psi(\sigma)$, also $t^* = \underline{\sigma} < \sigma$.

Mit Hilfe dieser Schrittweiten und des Gütekoeffizienten einer brauchbaren

Richtung können wir die *Verfahren charakterisieren.* Im weiteren bezeichnen wir $\{x^k\}$ als Prozeß *mit*

$$\begin{aligned}&\textit{schwacher Unterrelaxation} \leftrightarrow \underline{\sigma}_k \leqq t_k \leqq t_k^* \quad \forall k\\&\textit{vollständiger Relaxation} \leftrightarrow t_k = t_k^* \quad \forall k\\&\textit{starker Unterrelaxation} \leftrightarrow q_k \underline{\sigma}_k = t_k,\ 0 < q_k < 1, \quad \forall k\\&\sigma\textit{-Relaxation} \leftrightarrow t_k \in [\text{Min}\{\sigma_k, t_k^*\}, \text{Max}\{\sigma_k, t_k^*\}] \quad \forall k\end{aligned}$$

bzw. *als*

$$\textit{Quasigradientenprozeß} \leftrightarrow 0 < c \leqq c_k \leqq 1 \quad \forall k .$$

Abstiegsverfahren mit Unterrelaxation haben bei stark konvexem Zielfunktional die Eigenschaft

$$\lim_{k \to \infty} x^k = \bar{x} . \tag{3.21}$$

Denn es ist entweder $\sum \|x^{k+1} - x^k\| < \infty$. Dann folgt mit dem CAUCHY-Kriterium (3.21) aus

$$\|x^p - x^q\| \leqq \sum_{v=q}^{p-1} \|x^{v+1} - x^v\| \leqq \sum_{v=q}^{\infty} \|x^{v+1} - x^v\| \tag{3.22}$$

bei $p > q$, beides beliebige natürliche Zahlen.
Oder es ist $\sum \|x^{k+1} - x^k\| = \infty$. Dann folgt aus dem Lemma von ZOUTENDIJK [11] $\bar{x} = x^*$.

Wir wollen nun das Verhalten von Prozessen mit schwacher Unterrelaxation bzw. mit σ-Relaxation *in der Nähe der nichtoptimalen Singularität a* untersuchen und daraus Vorschriften für die Wahl der Richtungen ableiten.

Betrachten wir einen Punkt x^0, der nahe an a liegt, und eine in x^0 brauchbare Richtung h. Diese Richtung sei auch in a brauchbar (vgl. (3.1)). Dann läßt sich aus der Stetigkeit von $\nabla \bar{f}$ die Abschätzung

$$\langle \nabla \bar{f}(x^0), h \rangle - p \left\langle \frac{x^0 - a}{\|x^0 - a\|}, h \right\rangle - 2\bar{M} \langle x^0 - a, h \rangle < 0 , \tag{3.23}$$

also aus (3.16) $-2\langle x^0 - a, h \rangle \leqq \sigma \|h\|^2$ herleiten. Für $\langle x^0 - a, h \rangle < 0$ folgt damit $\|x^0 + \sigma h - a\| \geqq \|x^0 - a\|$, und $\sigma \leqq \underline{\sigma}$ ergibt sich aus (3.19). Für $\langle x^0 - a, h \rangle \geqq 0$ folgt sofort aus der Konvexität von $\|x\|$ und aus (3.19)

$$\|x^0 + \sigma h - a\| \geqq \|x^0 - a\|, \quad \sigma \leqq \underline{\sigma} . \tag{3.24}$$

Die Richtung des steilsten Abstieges $h = -\nabla f(x^0)$ muß dabei nicht immer in a brauchbar sein für x^0 nahe an a:

Beispiel 3: Es kann im Fall $\|\nabla \bar{f}(a)\| > p > \frac{1}{2}\sqrt{3}\, \|\nabla \bar{f}(a)\|$ für hinreichend kleinen Abstand $\|x^0 - a\|$ die Richtung $-\nabla \bar{f}(a) - p \frac{x^0 - a}{\|x^0 - a\|}$ so gewählt werden, daß sie in a nicht brauchbar ist:

$$\left\langle \nabla \bar{f}(a),\ \nabla \bar{f}(a) + p \frac{x^0 - a}{\|x^0 - a\|} \right\rangle < p \left\| \nabla \bar{f}(a) + \frac{x^0 - a}{\|x^0 - a\|} \right\| . \tag{3.25}$$

Wegen der Stetigkeit von $\nabla\bar{f}$ ist auch die Richtung $-\nabla f(x^0) = -\nabla\bar{f}(x^0) - p\frac{x^0 - a}{\|x^0 - a\|}$ in a nicht brauchbar (vgl. Abb. 2).

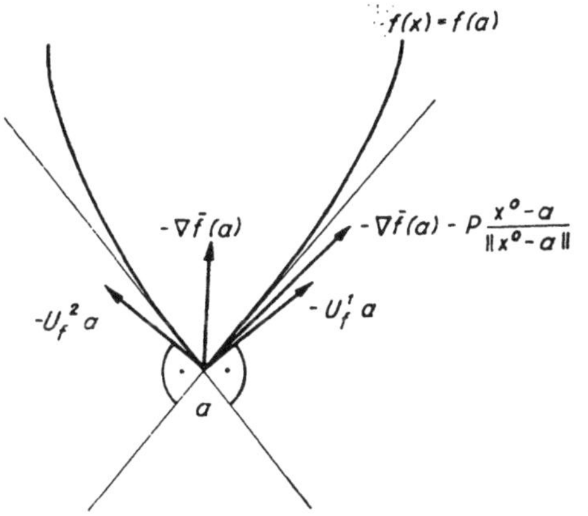

Abb. 2

Man kann aber auch längs der Richtung des steilsten Abstieges und längs kleiner Abweichungen dieser Richtung für x^0 nahe an a die Eigenschaft (3.24) zeigen (vgl. [3]). Diese Eigenschaft sagt aus, daß man von x^0 aus mit einem σ-Schritt oder einem schwachen Unterrelaxationsschritt längs einer auch in a brauchbaren Richtung oder längs einer fast optimal brauchbaren Richtung nicht näher an a gelangt. Daraus erhalten wir

Satz 3: *Sei $x^* \neq a$ und $\{x^k\}$ ein Abstiegsprozeß mit schwacher Unterrelaxation oder mit σ-Relaxation, seien für x^k nahe an a die Richtungen h^k in a brauchbar oder in x^k fast optimal brauchbar. Dann ist $\sigma_k \leqq \underline{\sigma}_k$ für x^k nahe an a, und es gilt*

$$\sum_{k=0}^{\infty} \|x^k - a\| = \infty . \tag{3.26}$$

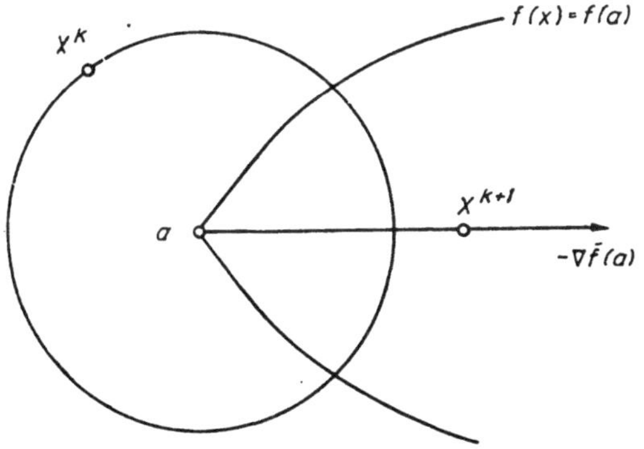

Abb. 3

Diese „abstoßende" Wirkung der nichtoptimalen singulären Stelle a wird für den Gradientenprozeß mit σ-Schrittweiten besonders deutlich. Und zwar erfolgt hier in erster Näherung das Fortschreiten längs $-\nabla\bar{f}(a)$, in Abhängigkeit vom Abstand zwischen x^k und a und nicht von der Lage des Punktes x^k (siehe Abb. 3).

Das folgt aus

$$x^{k+1} = x^k - \frac{1}{\overline{M}(x^k) + \frac{p}{\|x^k - a\|}} \nabla f(x^k) \tag{3.27}$$

$$= x^k - \frac{\|x^k - a\|}{p} \sum_{v=0}^{\infty} \left(- \frac{\overline{M}(x^k)}{p} \|x^k - a\| \right)^v \left(\nabla \bar{f}(x^k) + p \frac{x^k - a}{\|x^k - a\|} \right)$$

$$= a - \frac{\|x^k - a\|}{p} \nabla \bar{f}(a) + O(\|x^k - a\|^2).$$

Hieraus ergibt sich wegen $\|\nabla \bar{f}(a)\| > p$ für $\|x^k - a\| < \delta$ bei hinreichend kleinem $\delta > 0$

$$f(x^{k+1}) < f(a), \quad \|x^{k+n(k)} - a\| \geqq \delta \tag{3.28}$$

für eine natürliche Zahl $n(k) > 0$. Nach endlich vielen Schritten $n(k)$ verläßt also die Iterationsfolge die δ-Umgebung von a wieder und gelangt in das Innere der Niveaumenge $N_f(f(a))$.

Für Prozesse $\{x^k\}$ mit schwacher Unterrelaxation bzw. mit σ-Relaxation kann man des weiteren im Fall $\|\nabla \bar{f}(a)\| \neq p$ das Verhalten in der Nähe von a beschreiben mit:

$$\sum_{k=0}^{\infty} c_k^2 \|x^k - a\| < \infty, \quad \text{falls} \quad \lim_{k \to \infty} f(x^k) = f(a). \tag{3.29}$$

Dies erhalten wir daraus, daß wegen (3.15) und (2.5)

$$f(x^k) - f(x^{k+1}) \geqq \frac{\langle \nabla f(x^k), h^k \rangle^2}{2 \left(\overline{M}(x^k) + \frac{p}{\|x^k - a\|} \right) \|h^k\|^2} \tag{3.30}$$

$$\geqq \frac{\varepsilon^2}{2(\overline{M}M_0 + p)} c_k^2 \|x^k - a\|$$

gilt bei $\|x - a\| \leqq M_0$ für $x \in N_f(f(x^0))$. Die Summation über beide Seiten ergibt (3.29).

Die Divergenz der Reihe $\sum c_k^2 \|x^k - a\|$ schließt also für die betrachteten Prozesse die Konvergenz gegen a aus. Wir wollen nun *Konvergenzkriterien* herleiten. Dabei reicht es aus, die Konvergenz im Funktional, also $\lim f(x^k) = f(x^*)$ zu betrachten, da wegen der starken Konvexität hieraus die Konvergenz des Prozesses folgt: $\lim x^k = x^*$.

Bei stark und beschränkt konvexem Zielfunktional war die Divergenz der Reihe $\sum c_k^2$ eine notwendige und hinreichende Konvergenzbedingung für den Prozeß mit schwacher Unterrelaxation. In unserem Falle erhalten wir

Satz 4: *Ein Abstiegsverfahren mit schwacher Unterrelaxation bzw. mit σ-Relaxation konvergiert, wenn*

$$\sum_{k=0}^{\infty} c_k^2 \|x^k - a\| = \infty \tag{3.31}$$

gilt. Ist der Optimalpunkt regulär, dann ist diese Bedingung auch notwendig, und das Konvergenzverhalten wird im Einzugsbereich des Minimums beschrieben durch

$$f(x^k)-f(x^*) \leqq (f(x^{k_0})-f(x^*)) \prod_{j=k_0}^{k-1} \left(1-\frac{\bar{m}}{M} c_j^2\right) \tag{3.32}$$

bei

$$M=\bar{M}+\sup\left\{\frac{p}{\|x-a\|} \mid x \in N_f\left(\frac{1}{2} f(a)+\frac{1}{2} f(x^*)\right)\right\}.$$

Beweis: a) Hinlänglichkeit: Sei $\sum c_j^2 \|x^j-a\|=\infty$.
Mit (3.15) können wir die Konvergenzrate in Punkten $x^j \neq a$ folgendermaßen abschätzen:

$$\frac{f(x^{j+1})-f(x^*)}{f(x^j)-f(x^*)} \leqq 1-\frac{\|\nabla f(x^j)\|^2}{2\left(\bar{M}(x^j)+\frac{p}{\|x^j-a\|}\right)(f(x^j)-f(x^*))} c_j^2 . \tag{3.33}$$

Wegen (2.11), $\bar{M}(x^j) \leqq \bar{M}$ und $x^j \in N_f(f(x^0))$, also $\|x^j-a\| \leqq M_0$, schließen wir

$$\frac{f(x^{j+1})-f(x^*)}{f(x^j)-f(x^*)} \leqq 1-\frac{\bar{m}}{\bar{M}+\frac{p}{\|x^j-a\|}} c_j^2 \leqq 1-\frac{\bar{m}}{\bar{M} M_0+p} c_j^2 \|x^j-a\| . \tag{3.34}$$

Das ist auch für $x^j=a \neq x^*$ richtig. Wir bilden über beide Seiten das Produkt

$$\frac{f(x^k)-f(x^*)}{f(x^0)-f(x^*)} \leqq \prod_{j=0}^{k-1}\left(1-\frac{\bar{m}}{\bar{M} M_0+p} c_j^2 \|x^j-a\|\right). \tag{3.35}$$

Aus den Sätzen über konvergente Produkte (vgl. [7]) folgt mit $\sum c_j^2 \|x^j-a\|=\infty$ die Konvergenz des Prozesses.

b) Notwendigkeit: Sei $\lim x^k=x^*$ und $x^* \neq a$. Wäre $\sum c_j^2 \|x^j-a\|<\infty$, so wäre auch $\sum c_j^2<\infty$. Mit (2.7) und (2.12) haben wir für $j \geqq k_0$, also $x^j \in N_f\left(\frac{1}{2} f(a)+\frac{1}{2} f(x^*)\right)$

$$\frac{f(x^{j+1})-f(x^*)}{f(x^j)-f(x^*)} \geqq 1-\frac{\langle\nabla f(x^j), h^j\rangle^2}{2 \bar{m}\|h^j\|^2} \cdot \frac{2\left(\bar{M}+\frac{p}{\|x^j-a\|}\right)}{\|\nabla f(x^j)\|^2} \geqq 1-\frac{M}{\bar{m}} c_j^2 . \tag{3.36}$$

Wegen $\sum c_j^2<\infty$ ist $c_j \leqq \frac{\bar{m}}{M}$ für $j \geqq j_0 \geqq k_0$ und

$$\lim_{k \to \infty} \prod_{j=j_0}^{k}\left(1-\frac{M}{\bar{m}} c_j^2\right)>0 . \tag{3.37}$$

Mit

$$f(x^{k+1})-f(x^*) \geqq (f(x^{j_0})-f(x^*)) \prod_{j=j_0}^{k}\left(1-\frac{M}{\bar{m}} c_j^2\right) \tag{3.38}$$

ergibt das einen Widerspruch zu $\lim x^k=x^*$. Die Beschreibung des Konvergenzverhaltens ergibt sich aus (3.34).

Betrachten wir nun mit $\{x^k\}$ einen *Quasigradientenprozeß mit schwacher Unterrelaxation* oder *mit σ-Relaxation*. In Punkten x^k, die nahe an $a \neq x^*$ liegen, seien die Richtungen h^k in a brauchbar oder fast optimal brauchbar. Für das Beispiel 1 bedeutet die Forderung nach Brauchbarkeit der Richtungen h^k in a, daß die h^k zu K^+ gehören müssen. Aus Satz 3 folgt dann bei $0 < c \leqq c_k$ die Divergenz der Reihe $\sum c_k^2 \|x^k - a\|$, also die *Konvergenz* des Prozesses. Diese Prozesse konvergieren im Einzugsbereich des Minimums *linear* im Funktional. Es ist wegen (3.34)

$$\lim_{k\to\infty} \frac{f(x^{k+1}) - f(x^*)}{f(x^k) - f(x^*)} = \eta \leqq 1 - \frac{\bar{m}}{M} c^2 . \tag{3.39}$$

Liegt die singuläre Stelle a nahe am Optimalpunkt, dann beeinflußt sie die *Konvergenzgeschwindigkeit* des Prozesses in folgendem Maße:

$$\eta \leqq 1 - c^2 \frac{\bar{m}}{p} \left(1 - \frac{\|\nabla \bar{f}(a)\| - p}{p} \frac{\bar{M}}{\bar{m}}\right) \|x^* - a\| + O(\|x^* - a\|^3) . \tag{3.40}$$

Das erhalten wir, wenn wir auf (3.34) den oberen Limes anwenden und

$$\bar{m} \|x^* - a\|^2 \leqq \langle \nabla f(x^*) - U_f a, x^* - a \rangle \leqq (\|\nabla \bar{f}(a)\| - p) \|x^* - a\| \tag{3.41}$$

$$\text{bei} \quad U_f a = \nabla \bar{f}(a) - p \frac{\nabla \bar{f}(a)}{\|\nabla \bar{f}(a)\|}$$

ausnutzen. Je näher also x^* an a liegt, desto schlechter wird die Konvergenz. In der Grenze, wenn $\|\nabla \bar{f}(a)\| = p$ gilt, ist die Abschätzung (3.40) scharf in dem Sinn, daß für den konvergenten Gradientenprozeß mit σ-Schrittweiten wegen (3.27)

$$\lim_{k\to\infty} \frac{f(x^{k+1}) - f(a)}{f(x^k) - f(a)} = \lim_{k\to\infty} \frac{p \|x^{k+1} - a\| + \bar{f}(x^{k+1}) - \bar{f}(a)}{p \|x^k - a\| + \bar{f}(x^k) - \bar{f}(a)}$$

$$= \lim_{k\to\infty} \frac{\|x^{k+1} - a\|}{\|x^k - a\|} = \frac{\|\nabla \bar{f}(a)\|}{p} = 1 \tag{3.42}$$

gilt.

Die Notwendigkeit des Konvergenzkriteriums aus Satz 4 haben wir nur im Falle eines regulären Optimalpunktes nachgewiesen. Wenn a Optimalpunkt mit $\|\nabla \bar{f}(a)\| < p$ und $\{x^k\}$ ein Quasigradientenprozeß mit schwacher Unterrelaxation bzw. mit σ-Relaxation ist, so folgt nach (3.29) aus der Konvergenz des Abstiegsprozesses die Konvergenz der Reihe $\sum c_k^2 \|x_k - a\|$. Und umgekehrt folgt aus $\sum c_k^2 \|x^k - a\| < \infty$ wegen $0 < c \leqq c_k$ die Konvergenz des Prozesses. Für den betrachteten Abstiegsprozeß ist also $\sum c_k^2 \|x^k - a\| < \infty$ eine notwendige und hinreichende Konvergenzbedingung. Das zeigt, *daß das Konvergenzkriterium von der Regularität des Optimalpunktes abhängt.*

4. Anwendungen auf das Standortproblem

Beim Standortproblem ist die minimale Abstandssumme von n vorgegebenen festen Punkten a^i gesucht. Diese Aufgabe läßt sich im allgemeinen als ein Problem vom Typ (P) darstellen. Dabei nimmt a^1 mit dem minimalen Funktionswert unter den a^i die Stelle von a ein:

$$\sum_{i=2}^{n} p_i \|x-a^i\| + p_1 \|x-a^1\| = \underset{x\in R^2}{\mathrm{Min}!}\,. \tag{SP}$$

Diese *Abstandsfunktion* ist als Summe stetiger, konvexer Funktionen wieder stetig und konvex. Den Fall der Kollinearität von $a^2, \ldots, a^n$ schließen wir aus, da in diesem Fall

$$\bar{f}(x) = \sum_{i=2}^{n} p_i \|x-a^i\| \tag{4.1}$$

stückweise linear über der linearen Hülle der a^i, $2 \leqq i \leqq n$, ist.

Die Funktion $\bar{f}$ ist über $N_f(\alpha_0) = \{x \mid f(x) \leqq \alpha_0\}$ zweimal stetig differenzierbar mit

$$\langle \bar{H}(x)\, h, h\rangle = \lim_{t\to 0} \frac{\langle \nabla f(x+th) - \nabla f(x), h\rangle}{t} \tag{4.2}$$

$$= \sum_{i=2}^{n} \frac{p_i}{\|x-a^i\|} \sin^2 \sphericalangle (x-a^i, h) \cdot \|h\|^2 .$$

Die Matrix $\bar{H}$ der zweiten Ableitungen von $\bar{f}$ hat die Gestalt

$$H(x) = \sum_{i=2}^{n} \frac{p_i}{\|x-a^i\|} \left(E - \frac{(x-a^i)(x-a^i)^T}{\|x-a^i\|^2}\right), \quad E = \begin{pmatrix} 1 & 0 \\ 0 & 1 \end{pmatrix}, \tag{4.3}$$

mit den beiden Eigenwerten $\underline{\varrho}(x) \leqq \bar{\varrho}(x)$:

$$\underline{\varrho}(x) = \frac{1}{2} [\bar{\lambda}(x)^{-1} - A(x)], \quad \bar{\varrho}(x) = \frac{1}{2} [\bar{\lambda}(x)^{-1} + A(x)], \tag{4.4}$$

$$\underline{\varrho}(x) + \bar{\varrho}(x) = \bar{\lambda}(x)^{-1} = \mathrm{Spur}\, \bar{H}(x)$$

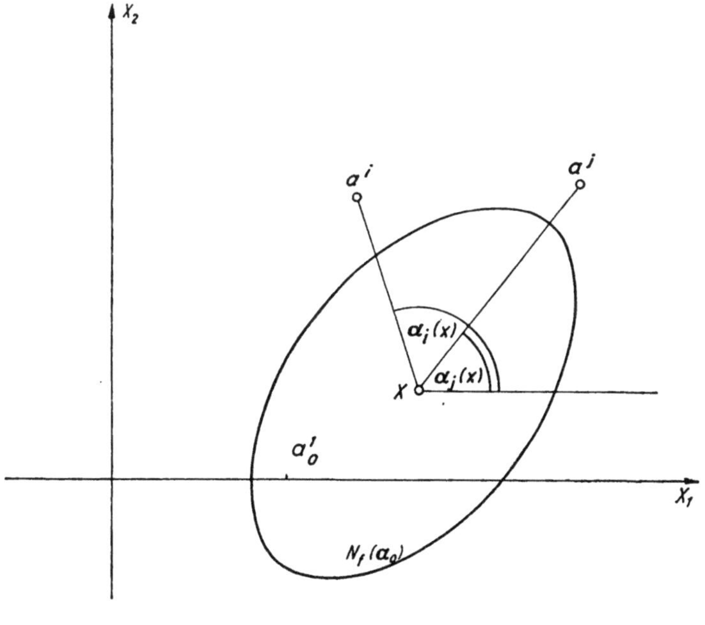

Abb. 4

bei

$$A(x)=\sqrt{\left(\sum_{i=2}^{n}\frac{p_i}{\|x-a^i\|}\sin 2\alpha_i(x)\right)^2+\left(\sum_{i=2}^{n}\frac{p_i}{\|x-a^i\|}\cos 2\alpha_i(x)\right)^2}\,, \tag{4.5}$$

$$\bar{\lambda}(x)^{-1}=\sum_{i=2}^{n}\frac{p_i}{\|x-a^i\|}\,.$$

Dabei bezeichnet $\alpha_i(x)$ den Winkel, den die Strecke $\overline{xa^i}$ mit der x_1-Achse einschließt (siehe Abb. 4).

Die Funktion $\bar{f}$ ist wegen

$$\langle\nabla\bar{f}\,(x+th)-\nabla\bar{f}(x),\,h\rangle=\int_0^1\langle\bar{H}\,(x+th)\,h,\,h\rangle\,\mathrm{d}t\leqq\int_0^1\bar{\varrho}\,(x+th)\,\|h\|^2\,\mathrm{d}t \tag{4.6}$$

über $N_f(\alpha_0)$ *beschränkt konvex* mit

$$\bar{M}_0=\sup\,\{\bar{\varrho}(x)\mid x\in N_f(\alpha_0)\}\,. \tag{4.7}$$

Um die starke Konvexität von $\bar{f}$ nachzuweisen, schätzen wir $\underline{\varrho}(x)$ über $N_f(\alpha_0)$ nach unten ab:

$$\underline{\varrho}(x)\geqq\frac{1}{2d_0}\left\{\sum_{i=2}^{n}p_i-\sqrt{\left(\sum_{i=2}^{n}p_i\sin 2\alpha_i\right)^2+\left(\sum_{i=2}^{n}p_i\cos 2\alpha_i\right)^2}\right\}, \tag{4.8}$$

wobei $d_0=\mathrm{Max}\,\{\|x-y\|\mid x,\,y\in N_f(\alpha_0)\}$ ist. Wir wenden auf den Ausdruck

$$\left(\sum_{i=2}^{n}p_i\sin 2\alpha_i\right)^2+\left(\sum_{i=2}^{n}p_i\cos 2\alpha_i\right)^2 \tag{4.9}$$

$$=\left(\sum_{i=2}^{n}p_i\right)^2-4\left(\sum_{i=2}^{n}p_i\cos^2\alpha_i\right)\left(\sum_{i=2}^{n}p_i\sin^2\alpha_i\right)+4\left(\sum_{i=2}^{n}p_i\sin\alpha_i\cdot\cos\alpha_i\right)^2$$

die Ungleichung $a-\sqrt{a^2-b^2}\geqq\frac{1}{2a}\,b^2$ an und erhalten

$$\underline{\varrho}(x)\geqq\frac{\left(\sum_{i=2}^{n}p_i\sin^2\alpha_i\right)\left(\sum_{i=2}^{n}p_i\cos^2\alpha_i\right)-\left(\sum_{i=2}^{n}p_i\sin\alpha_i\cos\alpha_i\right)^2}{d_0\cdot\sum_{i=2}^{n}p_i}\geqq 0\,. \tag{4.10}$$

Dabei trifft „$=0$“ nur dann zu, wenn

$$0=\sqrt{p_ip_j}\,\sin\,(\alpha_i(x)-\alpha_j(x)) \tag{4.11}$$

für jedes Paar (i,j) gilt. Wäre nun $\inf\,\{\underline{\varrho}(x)\mid x\in N_f(\alpha_0)\}=0$, dann würde wegen der Beschränktheit und Abgeschlossenheit von $N_f(\alpha_0)$ und wegen der Stetigkeit der $\alpha_i(x)$ ein $\bar{x}$ mit $\underline{\varrho}(\bar{x})=0$ existieren. Das widerspricht mit (4.11) dem, daß die a^i nicht kollinear sein sollten. Also ist $\bar{f}$ über $N_f(\alpha_0)$ *stark konvex* mit

$$\bar{m}_0=\inf\,\{\underline{\varrho}(x)\mid x\in N_f(\alpha_0)\}>0\,. \tag{4.12}$$

Der Kuhn*sche Algorithmus* [6], [10]

$$x^{k+1}=\left(\sum_{i=1}^{n}\frac{p_i}{\|x^k-a^i\|}\right)^{-1}\sum_{i=1}^{n}\frac{p_i}{\|x^k-a^i\|}a^i\quad\text{für}\quad x^k\neq a^i, \tag{4.13}$$

$$x^{k+1}=x^k\qquad\qquad\text{für}\quad x^k=a^i$$

ist ein Verfahren zur Lösung von (SP). KUHN interpretierte es als Schwerpunktverfahren und zeigte seine Konvergenz in Abhängigkeit vom Startpunkt x^0, da der Algorithmus an einer nichtoptimalen singulären Stelle a stoppen kann.

Wir interpretieren diesen Algorithmus als Gradientenverfahren und modifizieren die Iterationsvorschrift an Stellen $x^k = a^i$:

$$x^{k+1} = x^k - \lambda(x^k)\ \nabla f(x^k) \quad \text{bei} \quad \lambda(x^k) = \left(\sum_{i=1}^{n} \frac{p_i}{\|x^k - a^i\|}\right)^{-1} \tag{4.14}$$

$$x^{k+1} = a^1 - \underline{\sigma}(a^1)\ \nabla \bar{f}(a^1) \quad \text{für} \quad x^k = a^1.$$

Den Startpunkt x^0 wählen wir aus $N_f(\alpha_0)$.

Die Schrittweiten $\lambda(x^k)$ sind vom Typ σ_k mit $\bar{M}(x^k) = \bar{\lambda}(x^k)$. Dabei wählen wir $\bar{M} = \sup \{\bar{\lambda}(x) \mid x \in N_f(\alpha_0)\}$ als Schranke für die beschränkte Konvexität von $\bar{f}$. Hinreichend nahe am Optimalpunkt beschreiben diese Schrittweiten sogar starke Unterrelaxation. Und zwar gilt einerseits für $x^k \in N_f(\alpha_1)$, $f(x^*) < \alpha_1 < f(a^1)$,

$$\underline{\sigma}_k \geqq \frac{1}{\left(\bar{M}_1 + \frac{p_1}{2}\left(\frac{1}{\|x^k - \underline{\sigma}_k\ \nabla f(x^k) - a^1\|} + \frac{1}{\|x^k - a^1\|}\right)\right)} \geqq \frac{1}{M}, \tag{4.15}$$

wobei

$$M = \bar{M}_1 + \sup\left\{\frac{p_1}{\|x - a^1\|} \mid x \in N_f(\alpha_1)\right\}, \tag{4.16}$$

$$\bar{M}_1 = \sup\{\bar{\varrho}(x) \mid x \in N_f(\alpha_1)\}$$

sind. Andererseits sind $\lambda(x)^{-1}$, $\bar{\varrho}(x)$ und $\frac{p_1}{\|x - a^1\|}$ über $N_f(\alpha_1)$ stetig, also ist wegen (4.4) und (4.5)

$$\lambda(x)^{-1} > \lambda(x^*)^{-1} - \varepsilon = \frac{1}{2}\left(\lambda(x^*)^{-1} + \bar{\varrho}(x^*)\right) \tag{4.17}$$

$$\bar{\varrho}(x) + \frac{p_1}{\|x - a^1\|} < \bar{\varrho}(x^*) + \frac{p_1}{\|x^* - a^1\|} + \varepsilon = \frac{1}{2}\left(\lambda(x^*)^{-1} + \bar{\varrho}(x^*)\right)$$

mit

$$\varepsilon = \frac{1}{2}\left(\lambda(x^*)^{-1} - \bar{\varrho}(x^*) - \frac{p_1}{\|x^* - a^1\|}\right) > 0 \tag{4.18}$$

für alle $x \in N_f(\alpha_1)$, α_1 hinreichend klein. Daraus erhalten wir mit (4.15)

$$\lambda(x^k) < \frac{2}{\lambda(x^*)^{-1} + \bar{\varrho}(x^*)} \leqq \frac{1}{M} \leqq \underline{\sigma}_k \tag{4.19}$$

für $x^k \in N_f(\alpha_1)$. Damit haben wir gezeigt, daß der KUHNsche Algorithmus mit einer Modifizierung in $a^1 \neq x^*$ ein *Gradientenverfahren mit σ-Relaxation* ist und daß sich nahe am regulären Optimalpunkt *starke Unterrelaxation* einstellt.

Aus Satz 3 ergibt sich damit $\sum \|x^k - a^1\| = \infty$ für $x^* \neq a^1$. Da $c_k = 1$ für alle k ist, divergiert auch die Reihe $\sum c_k^2 \|x^k - a^1\|$. Für einen regulären Optimalpunkt ist damit die Konvergenz des Verfahrens gezeigt. Im Einzugsbereich des Minimums konvergiert es linear mit

$$\eta \leqq 1-\frac{\bar{m}}{\bar{M}}. \tag{4.20}$$

Liegt der Optimalpunkt x^* nahe an a^1, dann geht der Einfluß der Singularität auf die Konvergenzgeschwindigkeit wegen (3.19) aus

$$\eta \leqq 1-\frac{\bar{m}_0}{p_1}\left(1-\frac{\|\nabla \bar{f}(a^1)\|-p_1}{p_1}\frac{\bar{M}_0}{\bar{m}_0}\right)\|x^*-a^1\|+O\left(\|x^*-a^1\|^3\right) \tag{4.21}$$

hervor. Diese Abschätzung ist in der Grenze, $\|\nabla\bar{f}(a^1)\|=p_1$, scharf (vgl. [5]).

Ist a^1 Optimalpunkt mit $\|\nabla\bar{f}(a^1)\|<p_1$, so folgen Konvergenz und lineare Konvergenzgeschwindigkeit aus (3.42), die Reihe $\sum c_k^2\,\|x^k-a^1\|$ konvergiert.

Literatur

[1] Focke, J., A. Göpfert und H. Rudolph: Konvergenzuntersuchungen bei Gradientenverfahren. Math. Operationsforsch. Statist. **4** (1973) 1, 35–50.

[2] Göpfert, A.: Mathematische Optimierung in allgemeinen Vektorräumen. Teubner, Leipzig 1973.

[3] Illgen, A.: Das Verhalten von Abstiegsverfahren an Singularitäten des Gradienten mit Anwendung auf das Standortproblem. Dissertation, Leipzig 1976.

[4] Ioffe, A. D., i V. M. Tichomirov: Teorija ekstremal'nych zadač. Izdatel'stvo „Nauka", Moskva 1974.

[5] Katz, I. N.: Local convergence in Fermat's Problem. Mathematical programming **6** (1974), 89–104.

[6] Kuhn, H. W.: A note on Fermat's Problem. Mathematical Programming **4** (1973), 98–107.

[7] Knopp, K.: Theorie und Anwendung der unendlichen Reihen. Springer-Verlag, Berlin 1924.

[8] Ljubič, J. I., i G. D. Majstrovskij: Obščaja teorija relaksacionnych processov dlja vypuklych funkcionalov. Uspechi mat. nauk **1** (**151**) (1970), 57–109.

[9] Schumacher, K.: Untersuchungen zur Theorie der Gradientenverfahren. Dissertation, Tübingen 1974.

[10] Weiszfeld, E.: Sur le point pour laquel la somme des distances de n points donnés est minimum. Tôhoku Mathematical Journal **43** (1937) Juni.

[11] Zoutendijk, G.: Methods of feasible directions, a study in linear and nonlinear programming. Elsevier, Amsterdam 1960.

Резюме

Изучается поведение способов спуска для минимизации функций, которые почти всюду сильно и огранично выпуклы, кроме в окресности сингулярности градиента. Составляются в зависимости от параметров способа спуска необходимые и достаточные условия сходимости и указываются оценки скорости сходимости особенно в близости сингулярной точки. Этими средствами можно вообще доказать сходимость алгоритма Куна для проблемы Ферма в близости минимума.

Summary

In this paper the behavior of descent methods for minimization of functionals, which are in effect strong and bounded convex except in the neighbourhood of the singularity of the gradient is investigated. Sufficient and necessary conditions of convergence and assertions about the convergence speed in the neighbourhood of the singularity are stated in dependence of parameters of the descent methods. In general, in the "drainage area" of the minimum the convergence of the KUHN-Algorithm for solving the FERMAT-Problem may be proved with this means.

Eingereicht bei der Redaktion: 26. 1. 1978

H. A. Eiselt und H. von Frajer: **Operations Research Handbook. Standard Algorithms and Methods with Examples.** W. de Gruyter, Berlin, New York 1977, 398 Seiten, DM 74,–

Das vorliegende Handbuch enthält Standardalgorithmen, -methoden und -modelle der Operationsforschung. Das überdeckte Spektrum umfaßt: Lineare Optimierung, ganzzahlige Optimierung, Graphentheorie, Netzplantechnik, Spieltheorie, dynamische Optimierung, Bedienungstheorie, nichtlineare Optimierung, Simulation, Ersatzmodelle, Lagerhaltungsmodelle, Reihenfolgeprobleme und Probleme der optimalen Standortbestimmung. In einer schematischen Übersicht wird der Zusammenhang zwischen den behandelten Teilkomplexen deutlich gemacht. Als Benutzer werden von den Autoren anvisiert: Studenten der Ökonomie, der Sozialwissenschaften und der Mathematik sowie Praktiker der Operationsforschung. Die Arbeit mit dem Buch setzt die Kenntnis der Grundlagen von Differential- und Integralrechnung wie auch von Elementen der Wahrscheinlichkeitsrechnung voraus. Dem Buch vorangestellt ist eine Zusammenfassung der wichtigsten Begriffe und Definitionen aus der linearen Algebra, von denen Gebrauch gemacht wird. Zu loben sind Organisation und Darstellung des Stoffes. Wichtige bzw. häufig benutzte Methoden und Algorithmen werden in einheitlicher Form verständlich dargestellt. In jedem Abschnitt wird zunächst das jeweils betrachtete Problem formuliert. Darauf folgen in der Regel eine knappe Charakterisierung der der vorgeschlagenen Lösungsmethode zugrunde liegenden allgemeinen Konzeption sowie eine ausführliche Beschreibung des Verfahrens in einer den Programmablaufplänen angepaßten Art, was die „Anwenderfreundlichkeit" des Handbuchs erhöht.

Die getroffene Auswahl an Verfahren, Algorithmen und Modellen darf im großen und ganzen als dem Zweck des Buches entsprechend bezeichnet werden, wenngleich bei Neuauflagen in einigen Abschnitten Ergänzungen bzw. Änderungen zu wünschen wären, so hält beispielsweise der Rezensent im Abschnitt über nichtlineare Optimierung eine Darlegung des allgemeinen Konzepts der vielseitig anwendbaren Straf- und Barrieremethoden sowie der Lagrange-Multiplikatormethoden für nützlich. In Fällen, wo für dieselbe Grundaufgabe verschiedene Lösungsmethoden dargelegt werden, sollte ein Kommentar über spezifische Vorteile der Methoden nicht fehlen.

Für jede vorgestellte Methode wird ein einfaches Beispiel vollständig durchgerechnet. Dabei sind die Autoren im allgemeinen bemüht, anhand der Beispiele charakteristische Eigenheiten der Algorithmen und Modelle in Evidenz zu setzen. In einer Reihe von Fällen wäre ein deutlicher Hinweis auf zu erwartende Schwierigkeiten bei der praktischen Handhabung wünschenswert. Der Konzeption des Buches entsprechend, fehlen längere theoretische Erörterungen, Beweise und Hinweise auf Probleme bei der numerischen Realisierung der Verfahren. Es liegt in der Natur der Sache, daß die effektive Anwendung mancher Algorithmen zusätzliche Konsultationen der Spezialliteratur erfordert. Das Literaturverzeichnis will diesem Umstand Rechnung tragen. Die Auswahl entspricht wohl primär dem Geschmack der Verfasser. Die Ergänzung durch einige mathematisch orientierte Standardwerke und abschnittsweise Hinweise auf Bücher bzw. Artikel, in denen das mathematische Fundament der dargestellten Methoden und Modelle solide dargelegt wird, könnten den Gebrauchswert weiter erhöhen.

Insgesamt hebt sich das Handbuch vorteilhaft von einer Reihe ähnlich konzipierter Publikationen ab. Es erlaubt ein schnelles Kennenlernen einer repräsentativen Auswahl von Standardmethoden und -modellen der Operationsforschungspraxis und dürfte für den angesprochenen Interessentenkreis ein nützliches Hilfsmittel sein.

M. Schäuble

Math. Operationsforsch. Statist., Ser. Optimization, Vol. 10 (1979) No. 1, 57–65

Valuation of Some Random Search Methods[1]

CSILLA UGRAY[2]

Summary: The problem of finding a local minimum of a function $F(x)$, $x \in R^n$ is studied. The paper deals with some members of RASTRIGIN's random search family and with the method of random search with reversals. A new practical "Adaptive Step Size Random Search with Reversals" (ASSRSR) is proposed and analyzed. Computational experience is also reported. ASSRSR is compared with the adaptive step size random search algorithms of SCHUMER-STEIGLITZ and WHITE-DAY, and with the NEWTON-RAPHSON and FLETCHER-POWELL methods.

1. Introduction

The problem of locally minimizing a nonlinear function of several variables arises frequently in many areas of science. In recent years different techniques have been published to solve this type of problems.

Some methods require the gradient of the function to be calculated while others apply the approximations to the location of the minimum obtained in previous iterations. A third class of methods, which comprises the random search techniques, disregards both the function gradient and the previous function locations; steps are taken in random directions until a step is found which reduces the value of the function.

Random search methods for minimization were suggested earlier by BROOKS (1958) [1] and RASTRIGIN (1960, 1963, 1965) [3, 4, 6]. SCHUMER and STEIGLITZ analyzed the "Optimum Step Size Random Search" (OSSRS) [5], a theoretical model of the random search for the objective function

$$F(x) = \sum_{i=1}^{n} x_i^2$$

(x_i's are the components of vector x) and developed a practical algorithm "Adaptive Step Size Random Search" (ASSRS) [9] in 1968.

The criterion for comparison was the average number of necessary function evaluations for each method, in order to compute the minimum of the function

[1] The paper was presented at the conference „Mathematische Optimierung – Theorie und Anwendungen“ in Eisenach (GDR), December 5 to 10, 1977.

[2] Computing Center for Universities 1093, Budapest, Dimitrov tér 8. fszt. 28/K. Hungary.

with some specified accuracy. This criterion is one which is generally accepted in the field of function minimization.

An improved implementation of ASSRS was developed by WHITE and DAY in 1971 [12]. SCHRACK and BOROWSKI (1972) [7] tested the behaviour of ASSRS and other random methods with respect to the rate of convergence, dimensionality, sensitivity to noise.

The random search is of interest for several reasons: ease of programming, inexpensive realization, efficiency, flexibility, stability with respect to noise both inherent and generated.

An extension of RASTRIGIN's random search, called random search with reversals was investigated by SCHRACK and CHOIT (1976) [8]. This random search is theoretically superior to the above random searches.

Now a new practical adaptive step size random search with reversals (ASSRSR) [13] is proposed. The developed algorithm unifies the features of search direction reversing (after a failure) and adaptive step size random search procedures. It is shown that the average number of function evaluations required to minimize the functions

$$F_1(\boldsymbol{x}) = \sum_{i=1}^{n} x_i^2 \quad \text{and} \quad F_2(\boldsymbol{x}) = \sum_{i=1}^{n} a_i x_i^2$$

(a_i's are positive scalars) is an asymptotically linear function of their dimension. The performance of the ASSRSR algorithm is compared with the NEWTON-RAPHSON and FLETCHER-POWELL [2] methods.

2. The Step Size and Direction Problem

The discussion of this chapter follows the excellent description given by SCHRACK and CHOIT [8].

Let $F(\boldsymbol{x}), \boldsymbol{x} \in \boldsymbol{R}^n$, be an arbitrary real-valued function. The problem of finding a local minimum of $F(\boldsymbol{x})$ from a given approximation to the minimum $\boldsymbol{x}_0$ is studied. Sequential minimization methods can often be given by the formula

$$\boldsymbol{x}_{k+1} = \boldsymbol{x}_k + s_k \boldsymbol{u}_k, \quad k = 0, 1, \ldots, \tag{1}$$

where $\boldsymbol{x}$ and $\boldsymbol{u}$ are n-vectors, s is a scalar. The subscripts refer to the serial number of iterations. Assume the direction $\boldsymbol{u}_k$ is normalized to unit length so that the Euclidean norm $\|\boldsymbol{u}_k\| = 1$ for all k.

To define the algorithm (1) we have to deal with two fundamental problems: the choice of the step length and the direction.

RASTRIGIN [4] proposed and analyzed a sequential random search for the function $F_1(\boldsymbol{x})$ where unit random vectors distributed uniformly on the hypersphere are chosen as search directions. So the direction problem is eliminated.

Many proposals have been published to solve the step size problem: Fixed Step Size Random Search [4], OSSRS [5], ASSRS [9]. In the OSSRS the step length is the optimum at each step. The ASSRS algorithm attempts to adjust the step size to the optimum. For searches of this family, the probability of success [4, 5, 9] is

$$\mathsf{P}=\mathsf{P}\,(F_1(\boldsymbol{x}_{k+1})<F_1(\boldsymbol{x}_k))=\mathsf{P}(n,\varrho,s)=\mathsf{P}(n,t)$$

$$=\frac{\int_0^{\Phi_0}\sin^{n-2}\Phi\,\mathrm{d}\Phi}{2\int_0^{\pi/2}\sin^{n-2}\Phi\,\mathrm{d}\Phi}\,,\quad (\Phi_0=\arccos s/2\varrho)\,,$$

$t=s/\varrho$. (See Fig. 1 or [4, Fig. 3] or [9, Fig. 1] or [7, Fig. 1].) Note that if $t\to 0$ then $\mathsf{P}(n,t)\to 0.5$.

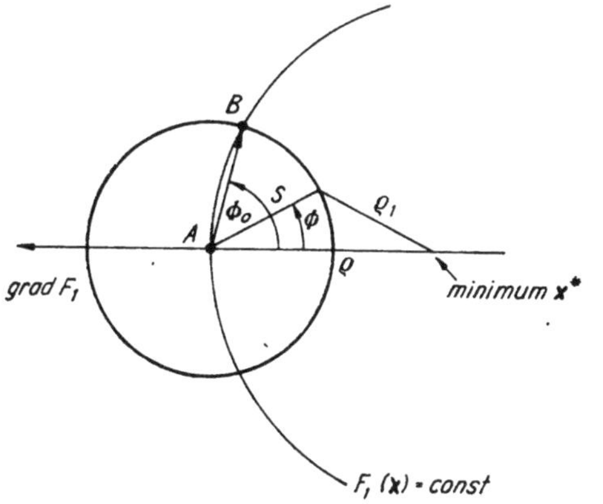

Fig. 1. $n=2$. A cross section of F_1

Φ is the angle between the direction of a random step and the negative gradient vector. The initial distance to the minimizing point is ϱ. A step of size s is taken at an angle Φ which reduces the distance from the minimum to ϱ_1. Φ_0 is the largest value of Φ giving an improvement of F_1.

3. Random Search Methods

3.1. Optimum Step Size Random Search Method (OSSRS)

is a mathematical model rather than an algorithm, it is not efficient on computer. The search is started at some point $\boldsymbol{v}_0$, the vector location of the initial approximation to the minimizing vector. In this method the following iteration is performed:

$$\boldsymbol{w}_k=\boldsymbol{v}_k+s_k\boldsymbol{u}_k,\quad k=0,1,\ldots,$$

$\boldsymbol{u}_k$'s are normalized random vectors, uniformly distributed on the sphere. The length s_k of the step is determined in order to maximize the expected percentage of reduction of F_1

$$\boldsymbol{v}_{k+1}=\begin{cases}\boldsymbol{w}_k, & \text{if}\quad F_1(\boldsymbol{w}_k)<F_1(\boldsymbol{v}_k)\,,\\ \boldsymbol{v}_k & \text{otherwise.}\end{cases}$$

The iteration is repeated until the desired accuracy is obtained. The expected percentage improvement I in $F_1(\boldsymbol{x})$ is given by

$$-\mathsf{E}(\Delta F_1)/F_1 = I(n, t) = \frac{\int_0^{\Phi_0} (2t\cos\Phi - t^2)\sin^{n-2}\Phi \, d\Phi}{2\int_0^{\pi/2} \sin^{n-2}\Phi \, d\Phi} \,. \tag{2}$$

To maximize $I(n, t)$, the right-hand side of (2) is differentiated with respect to t using LEIBNITZ's rule and set equal to zero. The following equation results for the optimal value of t:

$$\int_0^{\Phi_0} t \sin^{n-2}\Phi \, d\Phi = \frac{(1-t^2/4)^{\frac{n-1}{2}}}{n-1}\,, \quad 0 \leqq t \leqq 2 \,. \tag{3}$$

Upon making the appropriate approximations, the following asymptotic expressions are found for large n: (The assumption is made that for large n, t_{opt} is small, therefore $\sin^{-1} t_{\text{opt}}/2 \approx t_{\text{opt}}/2$. The right-hand side of (3) is approximated by the first three nonzero terms of its TAYLOR series expansion. STIRLING's approximation for $n!$ is used.)

$$t_{\text{opt}} = 1.225/\sqrt{n}, \quad I_{\max} = 0.406/n, \quad \mathsf{P}_{\text{opt}} = 0.27 \,.$$

(t_{opt} = the optimum relative step size, $I_{\max}$ = the maximum expected percentage improvement, P_{opt} = the optimum probability of improvement.)

It can be shown that the average number of function evaluations necessary to minimize F_1 to a prescribed accuracy is asymptotically a linear function of its dimension.

3.2. Adaptive Step Size Random Search (ASSRS)

is a feasible realization of OSSRS. An initial value s for the step size is chosen before each iteration. Then the iteration proceeds as follows: two random steps, one of size s and the other of size $s \cdot (1+b)$ are taken (b is a uniform pseudo-random number between 0 and 1). The step size which produces greater improvement in the function is chosen as the initial step size for the next iteration. If neither of the steps results in improvement, the step size remains unchanged and the iteration is repeated and if this occurs for p successive iterations, the step size is multiplied by a factor $0 < \delta < 1$. In addition, each time q iterations have passed, a step with the current step size is compared to a step $K > 1$ times as large.

3.3. Random Search with Reversals

is an extension of RASTRIGIN's random search. Let $\boldsymbol{z}$ be the parameter vector associated with the best function value previously found, i.e.

$$F(\boldsymbol{z}) \leqq F(\boldsymbol{x}_i), \quad i = 0, 1, \ldots, k \,.$$

After generating a random vector $\boldsymbol{u}_k$ and choosing the step size s_k, the next step can be calculated:

$$\boldsymbol{x}_{k+1} = \boldsymbol{z} + s_k \boldsymbol{u}_k$$

and the function value $F(\boldsymbol{x}_{k+1})$. Upon a first failure, the direction of search is reversed, i.e. $\boldsymbol{x}_{k+2}$ becomes:

$$\boldsymbol{x}_{k+2} = \boldsymbol{z} - s_k \boldsymbol{u}_k .$$

It is shown [8] that for fixed n and t:

$$\mathsf{P}_r(n, t) > \mathsf{P}(n, t), \quad n \geqq 2, \quad 0 \leqq t < 2 ,$$

where P_r and P is the probability of success of search technique with reversals and without reversals. Note that $\mathsf{P}_r \to 2/3$ if $t \to 0$. The analogous inequality

$$I_r(n, t) > I(n, t), \quad n \geqq 2, \quad 0 < t < 2 ,$$

holds for the respective expected relative improvements per function evaluations.

3.4. Adaptive Step Size Random Search with Reversals (ASSRSR)

A new practical algorithm ASSRSR [13] is proposed, which unifies the features of two methods: ASSRS and random search with reversals. The ASSRSR uses the parameters of White and Day [12]. A flow diagram for the ASSRSR method is given on Figure 2 (supposing that the theoretical optimum equals 0). Two random steps of size s and $s \cdot (1+a)$ are taken in random directions $\boldsymbol{u}_i$ and $\boldsymbol{up}_i$, where a is a uniform pseudo-random number between 0 and 1. The random directions are obtained by selecting random vectors uniformly distributed on the surface of the unit hypersphere.

4. Experimental Results of ASSRSR-Conclusion

The ASSRSR algorithm was programmed [11] in Algol-60 and tested on a Razdan-3 computer. The functions used were $F_1(\boldsymbol{x})$ and $F_2(\boldsymbol{x})$. Since near the minimum many functions can be validly approximated in a quadratic form, it is worth considering quadratic test functions. The test functions were minimized for $n = 5, 10, 15, 20, 25$ dimensions. For each case 4 independent trials were made. The procedure was initiated using a unit length random direction vector. The stopping criterion was $F < 10^{-10}$. The constants a_i were chosen on the interval (0, 1) by using a pseudo-random number generator. Now the value 10^{-10} is used as stopping criterion because this value means a very high accuracy on Razdan-3 computer. The plot of function calls required to minimize the function versus dimension was approximately linear in both cases. The slope of line was 88 for the function F_1 and 118 for F_2.

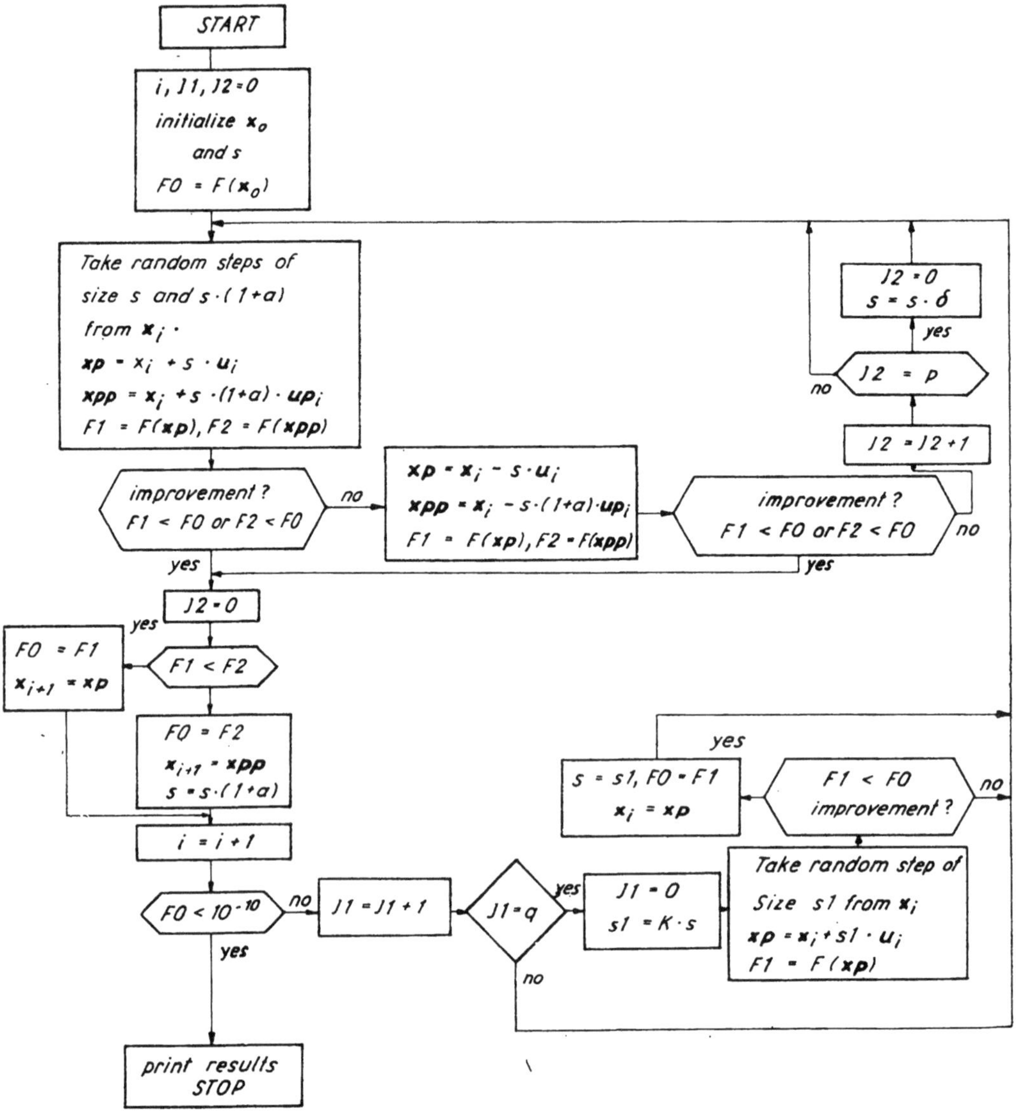

Fig. 2. Flow diagram for ASSRSR

To compare these results to those for ASSRS of SCHUMER-STEIGLITZ and WHITE-DAY, the slopes of our lines have to be adjusted to compensate for the different terminating values of F. (The terminating value of F for ASSRS was 10^{-8}.)

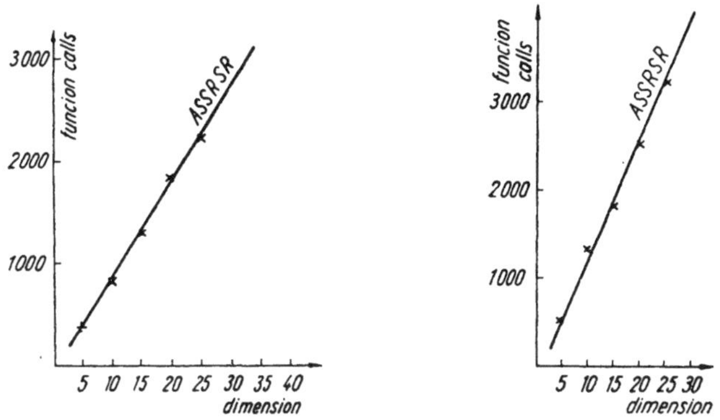

Fig. 3. Average number of function evaluations vs. dimension for $F_1(x)$. Stopping criterion $F_1 < 10^{-10}$

Fig. 4. Average number of function evaluations vs. dimension for $F_2(x)$. Stopping criterion $F_2 < 10^{-10}$

Comparison of ASSRSR with ASSRS of Schumer-Steiglitz and White-Day for, $F < 10^{-8}$

Method	Slope for F_1	Slope for F_2
ASSRS (Schumer-Steiglitz) [9]	80	84.2
ASSRS (White-Day) [12]	62	81.4
ASSRSR (this paper) [11, 13]	70.4	94.4

Consideration has to be given to the fact that the applied pseudo-random numbers (especially the selected a_i constants, moreover, the generated random directions) have effects on the results.

Comparison of the ASSRSR Method with the Newton-Raphson Method

It can be seen for the well-known deterministic method (the Newton-Raphson method) that the number of necessary function evaluations to locate the minimum depends quadratically on its dimension for F_1.

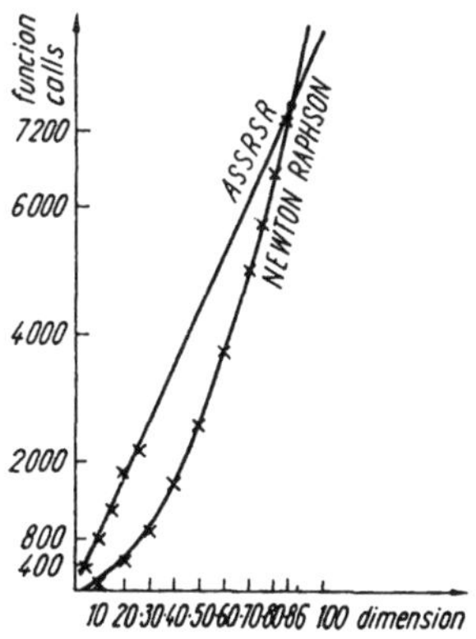

Fig. 5. Comparison of the ASSRSR and Newton-Raphson methods for F_1. Stopping criterion $F_1 < 10^{-10}$

Extrapolating the curve obtained for ASSRSR for the function F_1, the curves for the NEWTON-RAPHSON method and ASSRSR intersect at $n=86$ (for $F_1 < 10^{-10}$). Beyond this dimension, ASSRSR is superior.

Comparison of FLETCHER-POWELL and ASSRSR Methods

The FLETCHER-POWELL method [2] minimizes a function by means of conjugate directions. It can locate the minimum of a quadratic function of n variables in at most n iterations. A program called FLP [10] was written in the Algol-60 language to implement the FLETCHER-POWELL method on Razdan-3 computer. The functions F_1, F_2 were minimized for 20 to 140 dimensions, in steps of 20 dimensions. The starting point for each minimization was an n component vector whose elements were chosen by a pseudo-random number generator from the interval $(-0.5, 0.5)$. The vector was then normalized to unit length. The minimization was terminated when the objective function had been reduced to 10^{-10}. a_i's were randomly chosen from a uniform distribution on the interval (0, 1).

In both cases the results were well described by straight line functions whose slopes were about 8.4 for F_1 and 114.5 for F_2. In both cases the FLETCHER-POWELL algorithm was superior. For $F_1(\boldsymbol{x})$ a drastic difference can be seen between FLP and ASSRSR (Figure 6).

For $F_2(\boldsymbol{x})$ the approximating lines are very near (Figure 7). The better performance of FLP can be partly explained by the fact that FLP is especially suitable for quadratic problems. Note that in the case of F_2 (a far more frequently occuring approximation in practice) the difference is almost negligible.

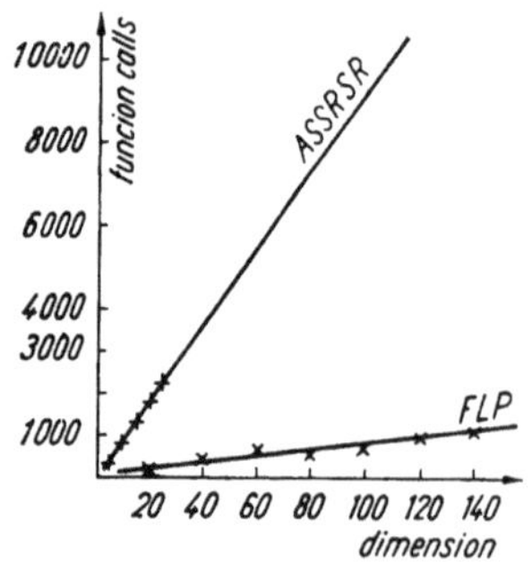

Fig. 6. Comparison of ASSRSR and FLP methods for F_1. Stopping criterion $F_1 < 10^{-10}$

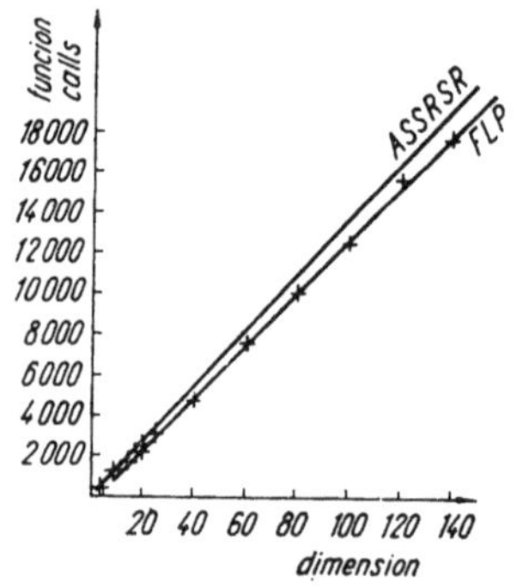

Fig. 7. Comparison of ASSRSR and FLP methods for F_2. Stopping criterion $F_2 < 10^{-10}$

References

[1] BROOKS, S. H.: Discussion of Random Methods for Locating Surface Maxima. Operations Research 6 (1958), 244–251.

[2] FLETCHER, R., and M. J. D. POWELL: A Rapidly Convergent Descent Method for Minimization. Computer Journal 6 (1963), 163–168.

[3] RASTRIGIN, L. A.: Extremal Control by the Method of Random Scanning. Automation and Remote Control 21 (1960), 891–896.
[4] RASTRIGIN, L. A.: The Convergence of the Random Search Method in the Extremal Control of a Many Parameter System. Automation and Remote Control. 24 (1963), 1337–1342.
[5] RASTRIGIN, L. A., and V. A. MUTSENIYEKS: Extremal Control of Continuous Multiparameter Systems by the Method of Random Search. Engrg. Cybernetics 2 (1964), 82–90.
[6] RASTRIGIN, L. A.: Optimization of Multivariate Systems by Random Search. Riga USSR, 1965.
[7] SCHRACK, G., and N. BOROWSKI: An Experimental Comparison of Three Random Searches. In: LOOTSMA, F. A. (Ed.): Numerical Methods for Nonlinear Optimization. Academic Press, London 1972, pp. 137–147.
[8] SCHRACK, G., and M. CHOIT: Optimized Relative Step Size Random Searches. Mathematical Programming 10 (1976), 230–244.
[9] SCHUMER, M. A., and K. STEIGLITZ: Adaptive Step Size Random Search. IEEE. Trans. Automat. Contr. AC-13 (1968) 3, 270–276.
[10] UGRAY, Cs.: Finding a Local Minimum of a Function of Several Variables by Fletcher-Powell Method. ESzK 10. Tájékoztató, pp. 29–32, 1973.
[11] UGRAY, Cs.: Finding a Local Minimum of a Function of Several Variables Applying Adaptive Step Size Random Search with Reversals. ESzK 21. Tájékoztató, 1978. (to be published)
[12] WHITE, L. J., and R. G. DAY: An Evaluation of Adaptive Step Size Random Search. IEEE. Trans. Automat. Contr., AC-16 (1971) 5, 475–478.
[13] UGRAY, Cs.: Some Random Search Function Minimization Techniques. ESzK 21. 1977. Budapest.

Zusammenfassung

In der vorliegenden Arbeit wird das Problem der Bestimmung eines lokalen Minimums der Funktion $F(\boldsymbol{x})$, $\boldsymbol{x} \in \boldsymbol{R}^n$, betrachtet. Der Artikel beschäftigt sich mit einigen Verfahren der Zufallssuche von RASTRIGIN und mit dem Verfahren der Zufallssuche mit Richtungsumkehr. Ein neues praktisches Verfahren der Zufallssuche mit adaptiver Schrittlänge und mit Richtungsumkehr (ASSRSR) wird dargelegt und analysiert. Rechenerfahrungen werden mitgeteilt. Das Verfahren ASSRSR wird mit den Methoden von NEWTON-RAPHSON, von FLETCHER-POWELL und auch mit den Zufallssuchverfahren der adaptiven Schrittlänge von SCHUMER-STEIGLITZ und von WHITE-DAY verglichen.

Резюме

В этой статье рассматривается поиск локального минимума функции многих переменных ($F(\boldsymbol{x})$, $\boldsymbol{x} \in \boldsymbol{R}^n$). Описываются некоторые методы случайного поиска, изложенные РАСТРИГИНЫМ и метод случайного поиска с изменением направления на противоположное. А также излагается новый метод случайного поиска адаптивного шага с изменением направления на противоположное (ASSRSR), приводятся результаты численной реализации этого метода. Дается сравнение алгоритма ASSRSR с методом случайного поиска адаптивного шага ШУМЕРА-ШТЕЙГЛИЦА и УАЙТА-ДЕЙА, и с методами НЬЮТОНА-РАФСОНА и ФЛЕТЧЕРА-ПАУЭЛЛА.

Eingereicht bei der Redaktion: 24. 1. 1978; revidierte Fassung: 25. 5. 1978

Book Review

J. Hoffmann-Jørgensen, T. M. Liggett, J. Neveu: **École d'Été de Probabilités de Saint-Flour VI-1976.** Edité par P.-L. Hennequin. Lecture Notes in Mathematics 598, Springer Verlag, Berlin-Heidelberg-New York 1977, 447 S., DM 37,–

Der vorliegende Band enthält drei umfangreiche, wichtige Arbeiten mit Übersichtscharakter und entsprechenden Literaturverzeichnissen, die auf der Grundlage von Vorträgen der Autoren auf der sechsten École d'Été de Probabilités in Saint-Flour 1977 gehalten wurden.

Zunächst stellt Hoffmann-Jørgensen („Probability in Banach space", S. 1–186) die Grundlagen der Wahrscheinlichkeitstheorie in Banachräumen dar, wobei es vor allem darum geht, die klassischen Grenzwertsätze auch für Banachraum-wertige Zufallsvariable zu beweisen. Dazu werden interessante mathematische Hilfsmittel benutzt, nämlich konvexe Maße (u. a. beim Beweis von Null-Eins-Gesetzen), spezielle geometrische Eigenschaften von Banachräumen und zwei Halbordnungsrelationen für Zufallsvektoren. Letztere spielen eine wichtige Rolle beim Beweis von Schranken für Normen, die für notwendige Bedingungen bei Grenzwertsätzen benötigt werden.

Es folgt der interessante und instruktiv geschriebene Aufsatz „The stochastic evolution of infinite systems of interacting particles" von T. M. Liggett (pp. 187–248). Hier werden zwei stochastische Modelle der Physik betrachtet, die seit etwa 10 Jahren mit Hilfe von Methoden aus der Theorie der Markowschen Prozesse behandelt werden, nämlich das spin-flip-Modell, das in allgemeiner Form zuerst Dobrushin (1971) studierte und der exclusion-Prozeß, der von Spitzer (1970) eingeführt wurde. Bei dem ersten Modell wird ein abzählbares System von Plätzen betrachtet, auf denen sich je ein Teilchen (Atom) befindet, das entweder im Zustand 0 oder 1 (positiver oder negativer Spin) ist. Geschwindigkeitsfunktionen regeln die Veränderung der Teilchenzustände in Abhängigkeit von der jeweiligen Konfiguration. Das zweite Modell beschreibt z. B. Gase. Hier befindet sich auf jedem Platz höchstens ein Teilchen, und Geschwindigkeitsfunktionen beschreiben die Übergänge der Teilchen von besetzten zu freien Plätzen. Für diese Prozesse wird zunächst die Frage nach der Existenz und Eindeutigkeit behandelt, wobei natürlich zusätzliche Bedingungen an die Geschwindigkeitsfunktionen erforderlich sind. Die interessanteste Frage ist die nach der Ergodizität und nach invarianten Verteilungen, herrührend vom statistisch-mechanischen Ursprung der Modelle. Diese Probleme werden in großer Allgemeinheit behandelt, wobei drei Techniken benutzt werden, nämlich die, gekoppelte Prozesse zu betrachten (um z. B. Vergleiche durchzuführen), endliche Hilfssysteme zu verwenden und die Monotonie der freien Energie des Systems auszunutzen.

Den Abschluß des Bandes bildet die Abhandlung „Processus ponctuels" von J. Nevenu (pp. 249–445). Hier wird im Bourbaki-Stil (was u. a. bedeutet: vollständige Beweise, gute Systematik, interessante Übungsaufgaben, aber: nur knappe Bemerkungen zum Inhalt der Begriffe und Resultate, fast kein direkter Bezug auf das umfangreiche Literaturverzeichnis) eine sehr schöne, klar geschriebene Darstellung der wichtigsten Ergebnisse der Theorie der Punktprozesse gegeben. In Kapitel 1 werden die Grundbegriffe der Theorie behandelt, ferner Existenzprobleme, Konvergenz von Verteilungsgesetzen und der Poisson-Prozeß. Das Kapitel 2 ist stationären Punktprozessen gewidmet, wobei die Palmsche Verteilung im Mittelpunkt steht, außerdem werden markierte Punktprozesse, Ergodizität und Spektralanalyse behandelt. In der stochastischen Geometrie auftretende Mengen werden als Punktprozesse in abstrakten Räumen interpretiert. Im 3. Kapitel schließlich wird die Methode der Martingale in der Theorie der Punktprozesse dargestellt.

D. Stoyan

Math. Operationsforsch. Statist., Ser. Optimization, Vol. 10 (1979) No. 1, 67–77

The NICHOLSON Principle in Discrete Dynamic Optimization

GERTRAUDE GIERTH[1] and JOHANNES TERNO[1]

Summary: This paper deals with the application of discrete dynamic optimization to problems with a fixed initial and final state. The basic concept of NICHOLSON demands an interpretation of the problem like an N-stage decision process with monotonous separability, unique stage transformation and unique inverse stage transformation.

The solution of this problem follows with simultaneous optimization in two directions—once in the process direction and otherwise in the opposite direction. The proposed algorithm gives an automatic selection rule for the direction of the following optimization step in order to realize the minimum of the number of all states to be considered with the presented concept. With it, an essential reduction of the storage requirement compared to the conventional dynamic optimization is given.

1. Introduction

Dynamic optimization is one of the basic techniques in the discrete optimization. But it is well-known that an increase in the feasible states produces an increase in the number of evaluations at each stage. This "problem of dimensionality" leads in many cases to an overflow of the computer memory. In our considerations we use the mostly underlying model in dynamic optimization—the so-called N-stage decision process with the initial state x_N and the final state x_0. The main idea in dynamic optimization is to compute an optimal solution of the decision process in a backward procedure from x_0, in a forward procedure from x_N, respectively, with the help of the recursive stage transformations. On the premises that a unique inverse stage transformation in every stage of the decision process exists, it is possible to interpret this process in opposite direction. TAHA [14] formulates the following question: When would either process arise? On the formulated premises the forward and the backward computations yield equivalent results. However, most of the authors use the backward computation for the explanation of dynamic optimization. We think that, on the above premises it is a matter of the point of view, i.e. what is backward and what is forward here.

The NICHOLSON *algorithm* [11] determines a shortest path in a weighted graph from a given start node to a given final node. This algorithm works simultaneously from the start node and from the final node. The forward (from the start

[1] TU Dresden, Sektion Mathematik, DDR–8027 Dresden, Mommsenstr. 13

node) and the backward (from the final node) computations meet one another between start and final node and the computations are finished when a corresponding optimality criterion is fulfilled.

This paper deals with the initial-final value problem in an N-stage decision process. On the premises of the existence of a unique inverse state transformation we apply the NICHOLSON concept, that is we realize simultaneously both, a recursive forward procedure from the initial state and a recursive backward procedure from the final state of the decision process. There is an automatic decision whether a step is made in the forward or in the backward procedure. Our concept realizes for some classes of problems a significant reduction in the high-speed memory requirement and in the computing time and gives aspects to more effective computational concepts. The general framework is applied to a special model consisting of a separable objective function subject to linear restrictions where all variables are integers. First computer experiences are reported.

2. The NICHOLSON algorithm for the determination of shortest paths

The algorithms by FORD, MOORE, DIJKSTRA, NICHOLSON and others [2], [11] are known to determine a shortest path from a given start node to a given final node in a weighted graph $G(X, V)$. In this graph, X denotes the set of all arcs, while V denotes the set of all nodes of the graph. The NICHOLSON algorithm realizes simultaneously a DIJKSTRA algorithm from the start node and an *"inverse"* DIJKSTRA algorithm from the final node. Here, the inverse DIJKSTRA algorithm is the usual DIJKSTRA algorithm working from the final node with the redirected arcs of the given graph. Thus, one step of the NICHOLSON algorithm consists either of one step of the DIJKSTRA algorithm or of one step of the inverse DIJKSTRA algorithm.

Let L_j be the computed instantaneous path length from the start node to the node v_j and, correspondingly let T_j be the computed instantaneous path length from the node v_j to the final node. We denote by A_n the set of the nodes for which in the step n of the algorithm the shortest path length from the start node is known and let E_n be the corresponding set of such nodes for which in the step n of the algorithm the shortest path length to the final node is known. The NICHOLSON algorithm has found a shortest path from the start to the final node when the following inequality

$$\min_{v_j \in V} (L_j + T_j) \leqq \min_{v_j \notin A_n} L_j + \min_{v_j \notin E_n} T_j$$

holds.

3. N-stage decision processes

In the usual manner [7] let an N-stage decision process be a process with N successively arranged stages where the output state of one stage is the input state of the following stage (s. Fig. 1).

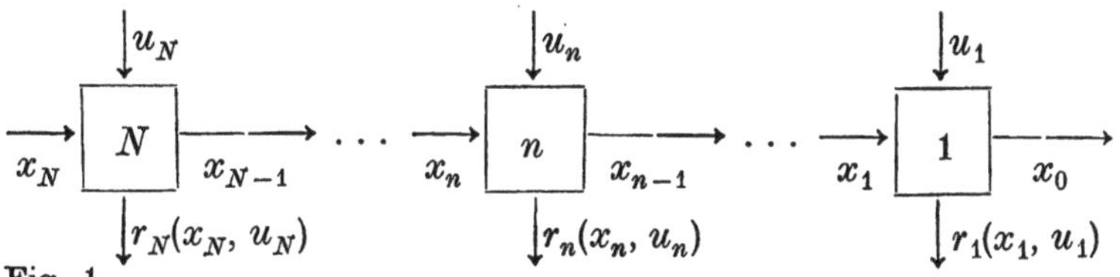

Fig. 1

In the stage n, $1 \leqq n \leqq N$, we obtain for a given input state x_n under the chosen decision u_n, the stage return $r_n(x_n, u_n)$ and the output state $x_{n-1} = t_n(x_n, u_n)$. The conversion from x_n to x_{n-1} is referred to as *state transformation*. x_N is the initial state of the decision process and x_0 the final one.

The application of the dynamic optimization concept requires a monotonous, separable total return R_N in dependence upon the stage returns in the sequence $r_N, r_{N-1}, \ldots, r_1$. Using the state transformation, we get for a fixed initial state x_N

$$R_N = h_N(r_N, r_{N-1}, \ldots, r_1) = R_N(x_N, u_N, \ldots, u_1) \tag{1}$$

and we assume the existence of

$$f_n(x_n) := \max_{u_n, \ldots, u_1} R_n(x_n, u_n, \ldots, u_1) \quad \text{for} \quad n = 1, \ldots, N.$$

Here $R_n(x_n, u_n, \ldots, u_1)$ is the total return for a process consisting of the stages $n, n-1, \ldots, 1$.

Supposing the existence of a unique inverse state transformation $x_n = \bar{t}_n(x_{n-1}, u_n)$ we can interpret the process in the opposite direction. That is we interpret x_{n-1} as the input state and x_n as the output state for the stage n (s. Fig. 2).

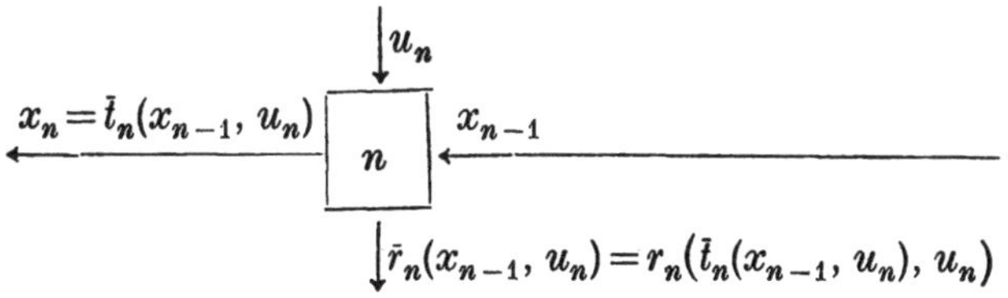

Fig. 2

If we apply the dynamic optimization to the opposite interpreted process we have to demand a monotonous separable dependence of the total return R_N from the stage returns $r_1, \ldots, r_N$ in this sequence.

Analogously to (1) we get for a fixed final state x_0

$$R_N = h_N(r_N, \ldots, r_1) = \bar{R}_N(u_N, \ldots, u_1, x_0) \tag{2}$$

where again we demand the existence of

$$\bar{f}_n(x_n) := \max_{u_N, u_{N-1}, \ldots, u_{n+1}} \bar{R}_{N-n}(u_N, \ldots, u_{n+1}, x_n)$$
$$\text{for} \quad n = N-1, N-2, \ldots, 0.$$

Here $\bar{R}_{N-n}(u_N, \ldots, u_{n+1}, x_n)$ is the total return for the subprocess consisting of the stages $N, N-1, \ldots, n+1$.

The objective is the determination of an optimal decision sequence what can be realized via the BELLMAN functional equations [1]:

$$f_1(x_1) = \max_{u_1 \in U_1(x_1)} r_1(x_1, u_1), \quad x_1 \in X_1 \Rightarrow u_1^*(x_1)$$

$$f_n(x_n) = \max_{u_n \in U_n(x_n)} \{r_n(x_n, u_n) \circ f_{n-1}(t_n(x_n, u_n))\}, x_n \in X_n \Rightarrow u_n^*(x_n), n = 2, \ldots, N,$$

where

$X_n = \{x_n\} :=$ set of feasible input states at stage n,
$U_n(x_n) :=$ set of feasible decisions at stage n under the input x_n,
$u_n^*(x_n) :=$ optimal decision at stage n under the input x_n,

and "$\circ$" denotes the monotonous, separable operator. That is, we decompose the decision process with the stages $n, n-1, \ldots, 1$ into the subprocess with the stages $n-1, \ldots, 1$ and the stage n (s. Fig. 3).

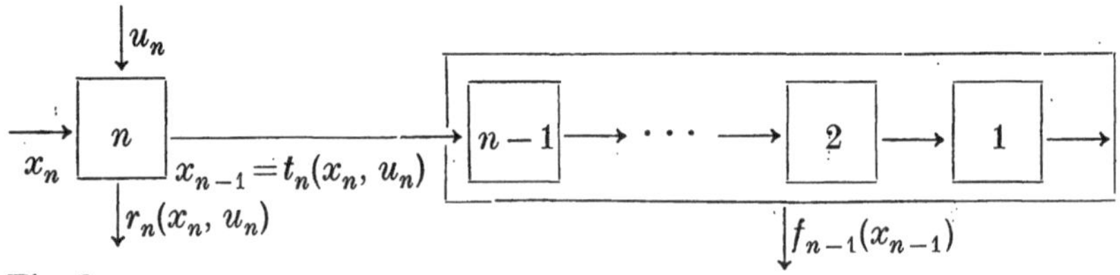

Fig. 3

The meaning of "$\circ$" can change from stage to stage. Analogously from (2) the BELLMAN recursion equations follow for the optimization in the opposite direction.

The application of the dynamic optimization to a given optimization problem

$$\Phi(y) = \max \quad \text{s.t.} \quad y \in M \tag{3}$$

requires the definition of stages with the corresponding returns and state transformations and especially the determination of the decision sets $U_n(x_n)$, $n = 1, \ldots, N$, and of the state sets X_n, $n = 1, \ldots, N$, on the base of the restrictions.

4. Application of the NICHOLSON principle to discrete dynamic optimization

4.1. Theoretical foundations

Let the initial state x_N and the final state x_0 for an N-stage decision process be given. Further we assume the validity of (1) and (2), that is especially, state transformations

$$x_{n-1} = t_n(x_n, u_n) \tag{4}$$

exist with unique inverse transformations

$$x_n = \bar{t}_n(x_{n-1}, u_n) \tag{5}$$

for $n=1, \ldots, N$. The decision variables u_n, $n=1, \ldots, N$, and the state variables take on discrete values only. Let be the composition operator "$\circ$" equal for all stages and restricted to addition or multiplication.

We define a *state transformation sequence* $t_{r,s}$, $r \geqq s$, leading via state x_r into state x_{s-1} using (4) with $u_n \in U_n(x_n)$ for $s \leqq n \leqq r$.

$$\text{Def. } 1: t_{r,s} = t_{r,s}(x_r, u_r, u_{r-1}, \ldots, u_s) := t_s(t_{s+1}[t_{s+2} \ldots t_{r-1}\langle t_r(x_r, u_r), u_{r-1}\rangle, \ldots, u_{s+1}], u_s) \quad 1 \leqq r, s \leqq N, r \geqq s.$$

For the inverse state transformations we similarly define an *inverse state transformation sequence* $\bar{t}_{r,s}$, $s \leqq r$, leading via state x_{s-1} into state x_r, using for this transformation (5) with $u_n \in \bar{U}_n(x_{n-1})$, $s \leqq n \leqq r$, where $\bar{U}_n(x_{n-1})$ denotes the set of feasible decisions at stage n under the input x_{n-1} for the redirected process.

$$\text{Def. } 2: \bar{t}_{s,r} = \bar{t}_{s,r}(x_{s-1}, u_s, u_{s+1}, \ldots, u_r) := \bar{t}_r(\bar{t}_{r-1}[\ldots \bar{t}_{s+1}\langle \bar{t}_s(x_{s-1}, u_s), u_{s+1}\rangle, \ldots, u_{r-1}], u_r) \quad 1 \leqq r, s \leqq N, s \leqq r.$$

To simplify the description we give

$$\text{Def. } 3: u_{r,s} := (u_r, u_{r-1}, \ldots, u_s), \; s \leqq r,$$

or $\quad u_{r,s} := (u_r, u_{r+1}, \ldots, u_s)$, $r \leqq s$, respectively, $1 \leqq r, s \leqq N$.

Now we choose an arbitrary stage k, $1 \leqq k \leqq N-1$. Using the state transformations $x_n = t_{n+1}(x_{n+1}, u_{n+1})$ of the stages N to $k+1$, or of the stages k to 1, respectively, we get, considering separability,

$$\begin{aligned} R_N &= R_{\text{II}}(x_N, u_N, u_{N-1}, \ldots, u_{k+1}) \circ R_{\text{I}}(x_k, u_k, u_{k-1}, \ldots, u_1) \\ &= R_{\text{II}}(x_N, u_{N,k+1}) \circ R_{\text{I}}(x_k, u_{k,1}), \quad x_k \in X_k. \end{aligned}$$

This precisely is the decomposition for which the principle of optimality is formulated.

With definitions 1 and 3 we can write

$$x_k = t_{N,k+1}(x_N, u_{N,k+1}) \quad \text{for every} \quad x_k \in X_k,$$

where X_k, $k=N-1, N-2, \ldots, 1$, is the set of states which can be reached from x_N using (4), without assuming to attain x_0. Now the initial-final state optimization problem can be written as follows:

$$g_{N,1}(x_N, x_0) := \max_{\substack{u_{N,k+1}, u_{k,1}, \\ x_k = t_{N,k+1}(x_N, u_{N,k+1}), \\ x_0 = t_{k,1}(x_k, u_{k,1})}} \{R_{\text{II}}(x_N, u_{N,k+1}) \circ R_{\text{I}}(x_k, u_{k,1})\} \tag{6}$$

where the restriction $x_0 = t_{k,1}(x_k, u_{k,1})$ reduces the set X_k to the set K of the exactly feasible states, what means

$$K := \{x_k \in X_k : x_k = t_{N,k+1}(x_N, u_{N,k+1}), x_0 = t_{k,1}(x_k, u_{k,1})\}\text{ [2]} \tag{7}$$

is the subset of all $x_k \in X_k$ from which x_0 can be reached, using (4) only.

Assuming the existence of the unique inverse state transformation (5) the set

[2] This definition corresponds to the definition of the feasible state set in [10], p. 44.

K can be described equivalently in the form as follows:

$$K = \{x_k \in X_k : x_k = t_{N,k+1}(x_N, u_{N,k+1}) = \bar{t}_{1,k}(x_0, u_{1,k})\} \\ = \{x_k^{(p)}, p = 1, \ldots, P\}.$$

We assume the finiteness of the set K.

Application of the monotonous separability immediately yields from (6)

$$g_{N,1}(x_N, x_0) = \max_{\substack{u_{N,k+1} \\ x_k = t_{N,k+1}(x_N, u_{N,k+1})}} \{R_{II}(x_N, u_{N,k+1}) \circ \max_{\substack{u_{k,1}, \\ x_0 = t_{k,1}(x_k, u_{k,1})}} R_I(x_k, u_{k,1})\}$$
$$= \max_{\substack{u_{N,k+1} \\ x_k = t_{N,k+1}(x_N, u_{N,k+1})}} \{R_{II}(x_N, u_{N,k+1}) \circ g_{k,1}(x_k, x_0)\}, \quad (8)$$

where $g_{k,1}(x_k, x_0)$ is the maximum total return of the subprocess consisting of the stages $k, k-1, \ldots, 1$ with the initial state x_k.

Consequently, the initial-final state optimization problem can be interpreted as a two-stage decision process (s. Fig. 4).

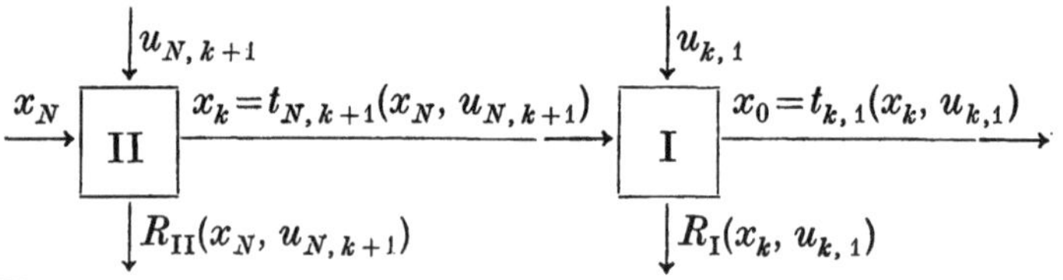

Fig. 4

Let be

$$U^p_{N,k+1} := \{u_{N,k+1} : t_{N,k+1}(x_N, u_{N,k+1}) = x_k^{(p)}\}, \quad p = 1, \ldots, P.$$

Therefore the state $x_k^{(p)} \in K \subset X_k$ can be reached with the feasible decision sequence $u_{N,k+1} \in U^p_{N,k+1}$. The states $x_k \in X_k \setminus K$ are of no interest for the following optimization as they do not lead to the final state x_0. Thus from (8) follows

$$g_{N,1}(x_N, x_0) = \max\{ \max_{u_{N,k+1} \in U^1_{N,k+1}} [R_{II}(x_N, u_{N,k+1}) \circ g_{k,1}(x_k^{(1)}, x_0)],$$
$$\max_{u_{N,k+1} \in U^2_{N,k+1}} [R_{II}(x_N, u_{N,k+1}) \circ g_{k,1}(x_k^{(2)}, x_0)], \ldots,$$
$$\max_{u_{N,k+1} \in U^P_{N,k+1}} [R_{II}(x_N, u_{N,k+1}) \circ g_{k,1}(x_k^{(P)}, x_0)]\}$$

and considering the monotonity

$$= \max\{[\max_{u_{N,k+1} \in U^1_{N,k+1}} R_{II}(x_N, u_{N,k+1})] \circ g_{k,1}(x_k^{(1)}, x_0), \ldots,$$
$$[\max_{u_{N,k+1} \in U^P_{N,k+1}} R_{II}(x_N, u_{N,k+1})] \circ g_{k,1}(x_k^{(P)}; x_0)\}$$
$$= \max_{x_k \in K} \{g_{N,k+1}(x_N, x_k) \circ g_{k,1}(x_k, x_0)\}, \quad (9)$$

where $g_{N,k+1}(x_N, x_k^{(p)}) = \max_{u_{N,k+1} \in U^p_{N,k+1}} R_{II}(x_N, u_{N,k+1})$ is the maximum total return of the subprocess consisting of the stages $N, \ldots, k+1$ with the initial state x_N and the final state $x_k^{(p)}$.

4.2. Application of the NICHOLSON principle

On the corresponding premises the initial-final value problem, formulated in Sec. 4.1, can be modelled in two ways. In the first way the process direction leads from x_N to x_0, whereas in the second, there is the process direction from x_0 to x_N. According to that we may use the BELLMAN recursion equations to optimize either from x_0 or from x_N. This and the theoretical foundations given in Sec. 4.1–especially formula (9)–enable us to extend the graph theoretical method described in Sec. 2–further on referred to as the NICHOLSON principle–to the problem mentioned previously.

We realize the essential simultaneous optimization from initial and final node in the following way. During the optimization from the so-called initial state x_N the state set X_{N-1}^N attainable in this direction is computed by the state transformation $x_{N-1} = t_N(x_N, u_N)$. Simultaneously we use the inverse state transformation to compute the state set X_1^0 attainable in the inverse direction from x_0. We want to minimize the number of states considered during the optimization of the N-stage process. Therefore if card $X_{N-1}^N \leqq$ card X_1^0, we carry out the first optimization step in the direction $x_N \to x_0$, otherwise in the direction $x_0 \to x_N$. Let be $X_j^N = \{x_j^N\}, j = N, N-1, \ldots$, the state set reached from x_N during the optimization in the direction $x_N \to x_0$ and let be $X_i^0 = \{x_i^0\}, i = 0, 1, 2, \ldots$, the corresponding sets for the inverse optimization direction $x_0 \to x_N$. Then we give the following automatic selection rule for the actual optimization direction after $r+s=t, t<N$, steps:

If $\sum_{j=N-r}^{N}$ card $X_j^N \leqq \sum_{i=0}^{s}$ card X_i^0, *then optimize one step in the direction* $x_N \to x_0$ *else optimize one step in the inverse direction* $x_0 \to x_N$.

The process ends after consideration of all stages, that is after $r+s=N$ steps. The choice of the optimization direction with the NICHOLSON principle yields a minimization of the number of considered states. We determine a stage k realizing this minimum:

$$\sum_{j=k+1}^{N} \text{card } X_j^N + \sum_{i=0}^{k} \text{card } X_i^0 := \min_{1 \leqq n \leqq N-1} \left[\sum_{j=n+1}^{N} \text{card } X_j^N + \sum_{i=0}^{n} \text{card } X_i^N \right].$$

Formula (9) in Sec. 4.1 contains a maximization of the total process comparing the maxima $g_{N,k+1}(x_N, x_k)$ from stage N to stage $k+1$ and $g_{k,1}(x_k, x_0)$ from stage k to stage 1 subject to all feasible states $x_k \in K$. It is not stated, how to compute the maxima $g_{N,k+1}(x_N, x_k)$ resp. $g_{k,1}(x_k, x_0), x_k \in K$. In the sense discussed above of the NICHOLSON principle optimal partial returns are computed both from x_N and from x_0, i.e. once $f_j^N(x_j^N)$, $x_j^N \in X_j^N$, is the optimal return for the stages N, $N-1, \ldots, j+1$ and other $f_i^0(x_i^0)$, $x_i^0 \in X_i^0$, is the optimal return for the stages 1, 2, $\ldots, i$. Therefore, reaching the stage k we do know the partial maxima as solutions of the initial-final state optimization problems $g_{N,k+1}(x_N, x_k^{(p)})$ resp. $g_{k,1}(x_k^{(p)}, x_0)$ for all $x_k^{(p)} \in K$. Hence (9) can be applied to determine the maximum.

In describing a general algorithm we use the following denotations:

$XN := \sum_{n=j}^{N} \text{card } X_n^N$ is the actual sum of the states reached from x_N,

$X0 := \sum_{n=0}^{i} \text{card } X_n^0$ is the actual sum of the states reached from x_0, $1 \leqq i \leqq j \leqq N$.

Algorithm:

Step 1: $XN := X0 := 1; j := N; i := 0; x_N^N := x_N; x_0^0 := x_0; f_{N+1}^N(x_N^N) := f_0^0(x_0^0) := 0;$

Step 2: If $j = i$ do $k := j$ and go to Step 4;
Determine X_{j-1}^N and card X_{j-1}^N applying the transformation $x_{j-1}^N = t_j(x_j^N, u_j)$, as X_{i+1}^0 and card X_{i+1}^0 applying the transformation $x_{i+1}^0 = \bar{t}_{i+1}(x_i^0, u_{i+1})$;

Step 3: If $XN + \text{card } X_{j-1}^N \leqq X0 + \text{card } X_{i+1}^0$ then $XN := XN + \text{card } X_{j-1}^N$, $j := j-1$ and compute for every $x_j^N \in X_j^N$ the optimal return $f_{j+1}^N(x_j^N) =$

$$= \max_{u_{j+1} \in \bar{U}_{j+1}(x_j^N)} \{r_{j+1}(\bar{t}_{j+1}(x_j^N, u_{j+1}), u_{j+1}) \bigcirc f_{j+2}^N(\bar{t}_{j+1}(x_j^N, u_{j+1}))\} \tag{10}$$

else $X0 := X0 + \text{card } X_{i+1}^0$, $i := i+1$ and compute for every $x_i^0 \in X_i^0$ the optimal return

$$f_i^0(x_i^0) = \max_{u_i \in U_i(x_i^0)} \{r_i(x_i^0, u_i) \bigcirc f_{i-1}^0(t_i(x_i^0, u_i))\}; \tag{11}$$

Go to Step 2;

Step 4: Determine $K := X_k^N \cap X_k^0$; if $K = \emptyset$ then there exists no feasible solution and the algorithm terminates;

Step 5: Calculate $x_k^{(p^*)}$ and $g_{N,1}(x_N, x_0)$ from

$$g_{N,1}(x_N, x_0) = \max_{x_k \in K} \{g_{N,k+1}(x_N, x_k) \bigcirc g_{k,1}(x_k, x_0)\}$$
$$= g_{N,k+1}(x_N, x_k^{(p^*)}) \bigcirc g_{k,1}(x_k^{(p^*)}, x_0);$$

Determine the optimal decision sequences $u_{k,1}^*$ and $u_{N,k+1}^*$ resp., the optimal total decision sequence $u_N^*, u_{N-1}^*, \ldots, u_1^*$ using the information stored during the solution of the problems (10) and (11).

4.3. Application to special problems and computational results

Let an integer optimization problem be written as follows:

$$\text{Maximize } F = \sum_{n=1}^{N} r_n(u_n) \tag{12}$$

$$\text{subject to } \sum_{n=1}^{N} a_{mn} u_n = b_m, \; m = 1, \ldots, M,$$

$$u_n \geqq 0, \text{ integer}, \; n = 1, \ldots, N,$$

where we assume $a_{mn}>0$, $b_m>0$, integer, $m=1, \ldots, M$, $n=1, \ldots, N$. The corresponding N-stage process is shown in Fig. 5. From the restrictions, the decision sets $U_n=\{u_n : 0 \leqq u_n \leqq \min_{1 \leqq m \leqq M} (b_m/a_{mn})\}$ are determined.

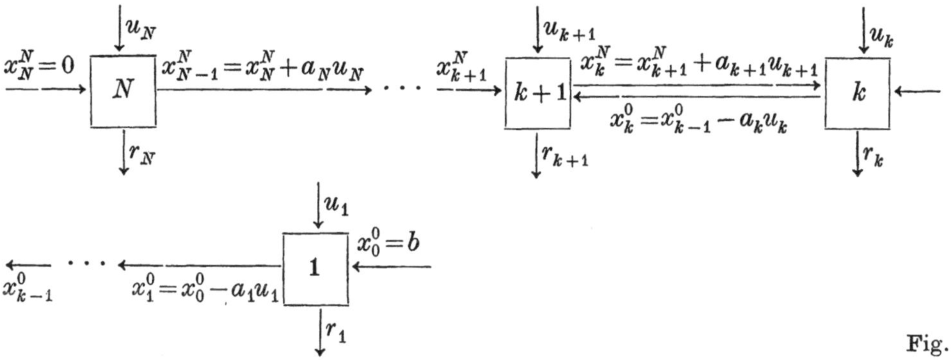

Fig. 5

With $x_n=(x_{1n}, x_{2n}, \ldots, x_{Mn})^{\mathsf{T}}$, $a_n=(a_{1n}, a_{2n}, \ldots, a_{Mn})^{\mathsf{T}}$, $b=(b_1, b_2, \ldots, b_M)^{\mathsf{T}}$ we get the unique transformations

$$x_{n-1}^N=t_n(x_n^N, u_n)=x_n^N+a_n u_n, \quad u_n \in U_n(x_n^N), \quad n=N, N-1, \ldots, 1,$$
$$x_n^0=\bar{t}_n(x_{n-1}^0, u_n)=x_{n-1}^0-a_n u_n, \quad u_n \in \bar{U}_n(x_{n-1}^0), \quad n=1, \ldots, N.$$

The constructed N-stage decision process possesses the required property of monotonous separability and this fact allows us to apply the described algorithm. This algorithm was coded in ALGOL. For some examples we compared the number of states to be considered in the conventional dynamic optimization—that means optimization in one direction only—and in the dynamic optimization with the NICHOLSON principle. Table 1 shows the results obtained for the given parameters N and M.

Table 1: Sum S of considered states and computer running time t

		Conventional dyn. opt.		NICHOLSON principle	
N	M	S	t sec.	S	t sec.
8	4	148	0.80	48	0.24
8	5	118	0.54	45	0.22
8	6	133	0.68	46	0.24
8	7	133	0.72	46	0.26
12	2	544	6.60	180	0.70
12	3	119	0.40	65	0.20
12	4	119	0.44	65	0.26
18	4	117	0.36	74	0.28
30	3	477	7.52	273	2.66

5. Conclusions

In the previous sections the application of the NICHOLSON principle to discrete dynamic optimization was dealt with. The underlying concept is based on the possibility of interpreting a given discrete optimization problem as an N-stage decision process with monotonous separability in which a unique state transformation with a unique inverse transformation exists. The numerical experiments in Sec. 4.3 show the essential reduction of the number of states to be considered in the dynamic optimization process. This reduction and further possibilities of a good realization of a NICHOLSON principle code allow to solve problems having larger numbers of variables and restrictions. It is hoped that within the framework of this paper some aspects are given to reduce the "problem of dimensionality", as it is called by R. BELLMAN.

6. References

[1] BELLMAN, R. E., and S. E. DREYFUS: Applied Dynamic Programming. Princeton University Press, Princeton, New Jersey 1962.

[2] DOMSCHKE, W.: Kürzeste Wege in Graphen: Algorithmen, Verfahrensvergleiche. Mathematical Systems in Economics 2, 1972.

[3] GESSNER, P., and H. WACKER: Dynamische Optimierung. Carl-Hanser-Verlag, München 1972.

[4] HADLEY, G.: Nonlinear and Dynamic Programming. Addison-Wesley, Reading, Mass. 1964.

[5] KORBUT, A. A., and J. J. FINKELSTEIN: Diskrete Optimierung. Akademie-Verlag, Berlin 1971.

[6] LARSON, R. E.: State Increment Dynamic Programming. American Elsevier Publishing Company, Inc., New York 1968.

[7] NEMHAUSER, G. L.: Einführung in die Praxis der dynamischen Programmierung. R. Oldenbourg-Verlag, München und Wien 1969.

[8] NEUMANN, K.: Reduzierung des Rechenaufwandes bei der Behandlung sequentieller Entscheidungsprozesse. Operations Research Verfahren—Methods of Operations Research VIII, 1970.

[9] — —: Dynamische Optimierung. BI – Hochschultaschenbücher Band 714, Bibliographisches Institut, Mannheim-Wien-Zürich 1969.

[10] — —: Operations Research Verfahren II. Carl-Hanser-Verlag, München-Wien 1977.

[11] NICHOLSON, T. A. J.: Finding the shortest route between two points in a network. Computer Journal 9 (1966), 275–280.

[12] PIEHLER, J.: Einführung in die Dynamische Optimierung. Teubner Verlagsgesellschaft, Leipzig 1966.

[13] SCHNEEWEISS, C.: Dynamisches Programmieren. Physica-Verlag, Würzburg-Wien 1974.

[14] TAHA, H. A.: Operations Research. Macmillan Publishing Co., Inc., New York, Collier Macmillan Publishers, London 1976.

[15] TERNO, J., G. GIERTH and H. WAGNER: Die Bestimmung des kürzesten Weges mittels dynamischer Optimierung. Wiss. Z. TU Dresden 1975.

Zusammenfassung

Die vorliegende Arbeit beschäftigt sich mit der Anwendung der diskreten Dynamischen Optimierung auf Probleme mit festem Anfangs- und Endzustand. Das zugrunde liegende Konzept von NICHOLSON verlangt die Möglichkeit einer Interpretation des Problems als

N-Stufen-Entscheidungsprozeß mit monotoner Separabilität, eindeutiger Stufentransformation, sowie eindeutiger inverser Stufentransformation. Die Lösung des Problems erfolgt durch gleichzeitige Optimierung in zwei Richungen – einmal in Prozeßrichtung und zum anderen in entgegengesetzter Richtung. Der vorgeschlagene Algorithmus steuert die Richtung, in der der folgende Optimierungsschritt durchzuführen ist, so daß die Gesamtzahl der betrachteten Zustände für diese Vorgehensweise minimal wird. Damit ergibt sich eine wesentliche Reduktion des Speicherplatzbedarfes gegenüber der herkömmlichen Dynamischen Optimierung.

Резюме

В статье рассматривается применение дискретной динамической оптимизации к проблемам с фиксированными начальным и конечным состояниями. Работа основана на принципе НИХОЛСЕНА. Это требует, что данную проблему можно интерпретировать как N-шаговый процесс с монотонной сепарабельностью, с однозначным шаговым преобразованием, и с однозначным обратным шаговым преобразованием. Решение проблемы производится путём одновременной оптимизацией по двум направлениям – во-первых по направлению протекания процесса и во-вторых по обратному направлению. Предлагаемый алгоритм выбирает направление, по которому должно проводиться следующий этап оптимизации, так, чтобы при этом минимизировать общее число рассмотренных состояний. Так значительно уменьшается необходимый обьём памяти машины по сравнению с обычной динамической оптимизацией.

Eingereicht bei der Redaktion: 29. 11. 1977

Book Review

M. Loève: **Probability Theory I.** Graduate Texts in Mathematics, vol. 45, Springer Verlag Berlin-Heidelberg-New York 1977, 4th edition, 425 S., DM 45,–

Dieses bewährte Werk liegt nunmehr in der 4. Auflage vor, zu Recht und zweibändig jetzt, da insbesondere die Brownsche Bewegung, Funktionalgrenzverteilungen und zufällige Irrfahrten neu aufgenommen wurden. Band I enthält zunächst einen Einleitungsteil (50 Seiten!), der die „elementare" Wahrscheinlichkeitstheorie umfaßt: intuitive Grundlagen; Axiome, Unabhängigkeit und den Bernoulli-Fall, einschließlich entsprechender spezieller Gesetze der großen Zahlen und Grenzwertsätze; Abhängigkeit und Markowsche Ketten mit diskretem Parameter. Sodann besteht der Band aus 3 Teilen: Grundlagen der Maßtheorie, allgemeine Konzepte und Werkzeuge der Wahrscheinlichkeitstheorie sowie Unabhängigkeit. Die notwendigen Grundbegriffe und -aussagen der Mengenlehre und der topologischen, metrischen und Banachräume, der additiven Mengenfunktionen, der Maßkonstruktion, der meßbaren Funktionen und Integration sowie der erforderlichen Konvergenzarten werden auf etwa 100 Seiten im ersten Teil streng eingeführt und bewiesen.

Im zweiten Teil (80 Seiten) wird von Wahrscheinlichkeitsräumen, Zufallsgrößen, zufälligen Vektoren und Funktionen ausgegangen, es werden Momente, Ungleichungen und Konvergenzen, Verteilungen, Verteilungsfunktionen, charakteristische Funktionen und deren Eigenschaften und Konvergenzen ausführlich behandelt.

Den größten Raum (170 Seiten) beansprucht Teil 3 unter dem Titel „Unabhängigkeit". Es werden die erforderlichen Erweiterungen des Unabhängigkeitsbegriffes vorgenommen und verschiedenartige Summen unabhängiger Zufallsgrößen untersucht, d. h., Konvergenz- und Stabilitätseigenschaften im Sinne der Gesetze der großen Zahlen werden bewiesen, auch für Summen von Zufallsgrößen, die bezüglich der Medianwerte zentriert bzw. beliebig symmetrisiert sind. Schranken werden angegeben. Alle klassischen Sätze vom Borelschen und Kolmogorowschen Null-Eins-Gesetz bis zum Gesetz vom iterierten Logarithmus werden bewiesen.

Die Problematik des zentralen Grenzwertsatzes wird mit allen klassischen und modernen Ergebnissen im historischen Ablauf der Abschwächung der Voraussetzungen vollständig entwickelt, wobei insbesondere deutlich gemacht wird, wie sich die für die einzelnen Etappen notwendigen Methoden (z. B. der charakteristischen Funktionen, des Stutzens und der Symmetrisierung) erweiterten.

Das letzte Kapitel dieses Teiles und damit des ersten Bandes überhaupt ist speziell Folgen unabhängiger und identisch verteilter Zufallsgrößen vorbehalten. Hierbei wird von der Fellerschen Darstellung Karamatas Theorie der regulären Variation ausgegangen und diese auf die Problematik der stabilen Verteilungsgesetze angewendet. Probleme von iid-Summen ohne Normierung, d. h. zufällige Irrfahrten, werden im abschließenden Paragraphen hinreichend ausführlich geklärt.

Der Band ist insgesamt, wie gesagt, ein sehr ausgereiftes Lehrwerk der genannten Teile der Wahrscheinlichkeitstheorie, in sich vollständig abgeschlossen, mathematisch exakt und didaktisch gekonnt für eine sehr solide Ausbildung von Studenten und anderen Interessenten verfaßt.

Der Leser ertrinkt nicht in der Fülle der bewiesenen Fakten, sondern genügend Zwischentexte sorgen für eine Hervorhebung des beweismethodischen Aspektes, häufig auch aus historischer Sicht, und geben auch knappe inhaltliche und Anwendungsmotivationen sowie Interpretationen. Die ausgezeichnete Aufmachung und drucktechnische Gestaltung tun gemeinsam mit den „Ergänzungen und Details" zu jedem Kapitel ein Übriges, um dieses Werk zu loben.

D. König

Math. Operationsforsch. Statist., Ser. Optimization, Vol. 10 (1979) No. 1, 79–83

Об использовании обобщенных множителей Лагранжа в комбинаторных методах дискретного программирования

К. Рихтер[1]

Резюме: В заметке устанавливается связь между предложенными С. С. Лебедевым [7–9] обобщенными множителями Лагранжа для задач дискретного программирования и множителями нелинейных функций Лагранжа. Далее сообщается некоторый вычислительный опыт решения дискретных задач с использованием множителей Лагранжа.

1. В [7–9] предложен подход к определению и использованию обобщенных множителей Лагранжа для задач целочисленного линейного программирования на базе введения дополнительного агрегирующего ограничения. В общем виде строится для прямой задачи

$$f = \min \{f(x) : x \in S,\ g(x) \leqq 0,\ p(x) \leqq 0\}\,, \tag{1}$$

где $f\colon R^n \to R^1$, $g\colon R^n \to R^m$, $p\colon R^n \to R^s$ и $S \subset R^n$ конечное множество, двойственная задача

$$L = \max \{L(\lambda) : \lambda \in R^m_+\}\,, \tag{2}$$

при чем

$$L(\lambda) = \min \{L(x, \lambda) : x \in S,\ v(x) \leqq 0\}\,, \tag{3}$$

$L(x, \lambda) = f(x) + (\lambda, g(x))$ функция Лагранжа, $v(x) = (v, p(x))$ и $v \in R^s_+$.

Если $p \equiv 0$, то задача (2) соответствует двойственным задачам в [3–6]. В таком виде задача (2) двойственна по обычным правилам построения двойственной задачи [3–6] к

$$f' = \min \{f(x) : x \in S,\ g(x) \leqq 0,\ v(x) \leqq 0\}\,. \tag{4}$$

В случае разрешимости задачи (1) имеется $f \geqq f' \geqq L$, т. е. L нижняя оценка значения f.

Данный прием позволяет иногда преодолеть разрыв двойственности. Характерной особенностью является то, что задачи (3) приводят к задачам рюкзачного типа. Для многих практических задач [1, 7–9] они распадают на мелкие подзадачи. Метод динамического программирования является эффективным методом их решения. Нелинейные, и в том числе кусочно-линейные

[1] Technische Hochschule Karl-Marx-Stadt, Sektion Mathematik, DDR – 90 Karl-Marx-Stadt, Reichenhainer Str. 41.

функции Лагранжа изучены многими авторами (см. [2]). Они позволяют преодолеть разрыв двойственности полностью. При этом соответствующие задачи (3) решаются так же трудно как исходная задача. Поэтому целесообразно использовать комбинацию линейной и нелинейной функций Лагранжа. На базе функции $L(x, \lambda, \mu) = L(x, \lambda) + \mu \cdot \max(0, v(x))$ двойственная к (4) задача принимает вид

$$L' = \max \{L(\lambda, \mu) : \lambda \in R^m_+, \mu \geqq 0\}, \tag{5}$$

где

$$\begin{aligned} L(\lambda, \mu) &= \min \{L(x, \lambda, \mu) : x \in S\} \\ &= \min \{L(x, \lambda) + \mu \cdot z : x \in S, v(x) - z \leqq 0, z \geqq 0\}. \end{aligned} \tag{6}$$

Следуя методике решения двойственных задач [3, 4, 6–9] оптимальное значение задачи (5) определяется числом

$$\begin{aligned} L' = \min \{ &\sum_{x \in S} v_x \cdot f(x) : \sum_{x \in S} v_x = 1, \sum_{x \in S} v_x \cdot g(x) \leqq 0, \\ &\sum_{x \in S} v_x \cdot \max(0, v(x)) \leqq 0, v_x \geqq 0, x \in S\}. \end{aligned}$$

Так как при $v(x) > 0$ обязательно выполняется $v_x = 0$, $x \in S$, то L' образуется только из значений аргументов задачи (3), т. е. $L = L'$. Тем самым предложенный С. С. Лебедевым подход к определению обобщенных множителей Лагранжа эквивалентен их определению на базе кусочно-линейных функций Лагранжа.

2. Успех решения задач дискретного программирования комбинаторными алгоритмами зависит в большой мере от того, насколько удается раскрыть специфику задач и приспосабливать к ним известные принципы решения. Для примера обсуждаем опыт решения одной динамической многопродуктовой задачи управления запасами [10] лексикографическим методом ветвей и границ. При обозначениях

$$\hat{B}_{jt} = \left\{ B_{jt}, B_{jt} + B_{jt+1}, \ldots, \sum_{s=t}^{T} B_{js} \right\}, \quad j \in M, \quad t = 1(1)\, T,$$

рассматриваем задачу

$$\begin{aligned} f = \min \Bigg\{ &\sum_{j \in M} \sum_{t=1}^{T} (d_j \cdot \operatorname{sign} x_{jt} + L_j y_{jt}) : y_{jt} = y_{jt-1} + x_{jt} - B_{jt} \geqq 0, \\ &x_{jt} \in \hat{B}_{jt} \text{ при } y_{jt-1} = 0, \quad x_{jt} = 0 \text{ при } y_{jt-1} > 0, \quad y_{j0} = y_{jT} = 0, \\ &j \in M, \quad \sum_{j \in M} b_j y_{jt} \leqq A_t, \quad t = 1(1)\, T \Bigg\}, \end{aligned} \tag{7}$$

где

$$d_j, L_j, b_j, B_{jt}, A_t \geqq 0, \quad j \in M, \quad t = 1(1)\, T.$$

Алгоритм решения задачи и ее экономическая интерпретация изложены в [10]. В обозначениях задачи (1) можно понимать под g функции $\sum_{j \in M} b_j y_{jt} - A_t$, $t = 1(1)\, T$, и включить в множество S все остальные ограничения. Функция p пологается нулевой. Если построить двойственную задачу типа

(2) для (7), то задачи (3) принимают вид

$$L(\lambda)=\sum_{j\in M}\min\left\{\sum_{t=1}^{T}(d_j\cdot\operatorname{sign}x_{jt}+(L_j+b_j\cdot\lambda_t)\cdot y_{jt}) : y_{j0}=y_{jT}=0\,,\right. \quad (8)$$

$$y_{jt}=y_{jt-1}+x_{jt}-B_{jt}\geqq 0,\quad x_{jt}\in\tilde{B}_{jt}\ \text{при}\ y_{jt-1}=0,\quad x_{jt}=0$$

$$\left.\text{при}\quad y_{jt-1}>0,\quad t=1(1)\,T\right\}-\sum_{t=1}^{T}\lambda_t\cdot A_t\,,$$

т. е. распадают на независимые однопродуктовые задачи, которые легко решаются методом динамического программирования [11].

Двойственная задача может быть решена методом генерации столбцов. В алгоритме [10] в качестве нижней оценки использовано $L(0)$. Такой подход эффективен, когда общие ограничения $g(x)\leqq 0$ не жестки, что равноценно тому, что $|f-L(0)|$ не велико. Представляет интерес, насколько применение улучшенной оценки L приведет к снижению времени счета в алгоритме [10]. Отметим следующие особенности:

а) В [1] наблюдается, что в некотором смысле наилучшая оценка L по всей допустимой области задачи (7) может быть не наилучшей в некоторых допустимых подобластях, и в том числе хуже тривиальной оценки по $L(0)$. Так как $L(0)$ определяется легко, то рекомендуется брать в качестве оценки максимум обеих оценок.

б) Сочетание лексикографического метода ветвей и границ с решением задач (8) методом динамического программирования имеет то преимущество, что оценки пересчитываются относительно просто при ветвлении. Пусть однопродуктовые задачи в (8) решаются следующей процедурой:

$$f_{jt}(\lambda)=\min\left\{f_{jts}(\lambda)+f_{js+1}(\lambda): s=t(1)\,T\right\}, \quad (9)$$

где

$$f_{jts}(\lambda)=d_j\cdot\operatorname{sign}B_{jts}+\sum_{l=t}^{s}(L_j+b_j\cdot\lambda_l)\cdot B_{j,l+1,s}\,,$$

$$B_{jts}=\sum_{l=t}^{s}B_{jl},\quad 1\leqq t\leqq s\leqq T,\quad f_{jT+1}(\lambda)\equiv 0,\quad j\in M,\quad \lambda\in R_+^T\,.$$

Тогда

$$L(\lambda)=\sum_{j\in M}f_{j1}(\lambda)-\sum_{t=1}\lambda_t\cdot A_t\,.$$

Если для некоторой допустимой подобласти V нижняя оценка обозначена через $p(\lambda)$, $y_{jt-1}=0$ и переменные x_{js}, y_{js}, $s=t(1)\,T$, не определены для некоторого $j\in M$, то оценка для $V'\subset V$, где $x_{jt}=B_{jtr}$, $x_{js}=0$, $s=t+1(1)\,r$, $y_{js}=$ $=B_{j,s+1,r}$, $s=t(1)\,r$, равна

$$p'(\lambda)=p(\lambda)+d_j\cdot\operatorname{sign}B_{jtr}+\sum_{s=t}^{r}(L_j+b_j\cdot\lambda_s)\cdot B_{j,s+1,r}+f_{j,r+1}(\lambda)-f_{j,t}(\lambda). \quad (10)$$

в) Алгоритм [10] отдельно и вместе с методом генерации столбцов и способом пересчета оценок (10) запрограммированы на алгоритмическом языке

Басик-ФОРТРАН и реализованы на ЭВМ ЕС 1020. Пусть t_G время счета для решения двойственной задачи методом генерации столбцов, t_0 время счета алгоритма [10] и t_L время счета алгоритма [10] с использованием приема пересчета оценок (10) и алгоритма решения двойственной задачи. Получены следующие результаты:

Параметры	Размеры задач ($\|M\|$, T)		
	(3,4)	(6,6)	(9,6)
f	41	264	384
L	38,7	260	370
$L(0)$	38	237	324
t_L в сек.	5	108	228
t_G в сек.	4	29	71
t_0 в сек.	1	49	937

Анализ расчетов позволяет сделать следующие выводы:

1) Использование обобщенных множителей Лагранжа в комбинаторных алгоритмах не всегда эффективнее работы с тривиальными оценками.

2) При жестких ограничениях $g(x) \leqq 0$ относительно $x \in S$ использование обобщенных множителей Лагранжа может быть существенно эффективнее работы с тривиальными оценками. В заключении следует отметить, что перспективным является решение двойственных задач дискретного программирования субградиентными методами [5], [12].

Литература

[1] Азарян, Л. Л., С. С. Лебедев и Л. М. Местецкий: Решение целочисленных задач транспортного типа с использованием обобщенных множителей Лагранжа. Экон. и мат. методы **XIII** (1977) 4, 723–731.

[2] Bazaraa, M. S.: Geometry and resolution of duality gaps. Nav. Res. Log. Quart. **20** (1973) 2, 357–365.

[3] Brooks, R., and A. M. Geoffrion: Finding Everett's Lagrange multipliers by linear programming. Oper. Research **14** (1966), 1149–1153.

[4] Fisher, M. L.: Optimal solution of scheduling problems using Lagrange multipliers. Part I. Oper. Research **21** (1973) 5, 1114–1127.

[5] Fisher, M. L.: A dual algorithm for the one-machine scheduling problem. Math. Programming **11** (1976), 229–251.

[6] Lasdon, L. S.: Optimization theory for large systems. London, MacMillan (1970).

[7] Лебедев, С. С.: Целочисленное программирование и множители Лагранжа. Экон. и мат. методы **X** (1974) 3, 592–610.

[8] Лебедев, С. С.: Использование обобшенных разрешающих множителей для решения задач целочисленного линейного программирования методом упорядоченного перебора. Мат. методы в экон. исследованиях, Наука (1974), 70–78.

[9] Лебедев, С. С.: Обобщенные множители Лагранжа и оценочные задачи в целочисленном программировании. Труды 6-й Зимн. школы по мат. прогр. и смежн. вопр., Дрогобыч 1973, Москва (1975), 114–157.

[10] Richter, K.: Deterministische dynamische Mehrproduktmodelle der Produktionsplanung und Lagerhaltung. Math. Operationsforschung u. Stat. **6** (1975) 3, 385–395.

[11] WAGNER, H. M.: Principles of Operations Research. Prentice-Hall, Inc. Englewood Cliffs, N. J. (1969).
[12] Шор, Н. З.: Обобщенные градиентные методы минимизации негладких функций и их применение к задачам математического программирования. Экон. и мат. методы 12 (1976) 2, 337–356.

Zusammenfassung

In der Notiz werden Beziehungen zwischen den LEBEDJEVschen verallgemeinerten LAGRANGEmultiplikatoren für Aufgaben der diskreten Optimierung und Multiplikatoren für nichtlineare LAGRANGEfunktionen hergestellt. Weiter wird über gewisse numerische Erfahrungen der Lösung diskreter Optimierungsaufgaben unter der Verwendung von LAGRANGEmultiplikatoren berichtet.

Summary

In this note relations between LEBEDJEV's generalized LAGRANGE multipliers for discrete programs and multipliers of nonlinear LAGRANGE functions are developed. Furthermore, some numerical experiences in solving discrete programs by LAGRANGE multipliers are discussed.

Eingereicht bei der Redaktion: 10. 1. 1978

Math. Operationsforsch. Statist., Ser. Optimization, Vol. 10 (1979) No. 1, 85–92

Ein Produktionsplanungsmodell bei horizontaler und vertikaler Verflechtung

K. Beer[1], J. Käschel[1] und K. Richter[1]

Zusammenfassung: Es wird ein Modell zur Erstellung eines optimalen Produktionsplanes für mehrere Produkte, Betriebsteile, Anlagen und Produktionsstufen mit gewissen Diskretheitsforderungen untersucht. Wenn die Anzahl der Produkte wesentlich größer als die Anzahl der Anlagen aller Betriebsteile und Stufen ist, so liefert ein zugeordnetes zweifach duales lineares Optimierungsproblem eine praktikable Lösung für die Ausgangsaufgabe. Einige rechentechnische und ökonomische Ergebnisse bei der Anwendung des Modells werden vorgestellt.

1. Aufgabenstellung

Wir stellen ein Optimierungsmodell vor, in dem sowohl die vertikale Verflechtung der Zwischen- und Endprodukte eines Betriebes als auch die horizontale (territoriale) Verflechtung der einzelnen, räumlich voneinander getrennten Betriebsteile Berücksichtigung finden. Die hier vorgestellte Version des Problems stellt eine gewisse Idealisierung des bei praktischen Berechnungen verwendeten Modells dar. Auf die konkreten Abweichungen dieser Variante von der real benutzten gehen wir verbal in der Bemerkung 1 ein.

Wir betrachten einen Betrieb als ein System von $|I|$ Teilsystemen oder Betriebsteilen. Der Betrieb fertigt eine Menge J von Endprodukten. Die Produktion dieser Erzeugnisse erfolgt in $|R|$ Stufen, wobei die Stufe $\{1\} \in R$ die Komplettierung der in den Stufen $r \in R - \{1\}$ hergestellten Halbfabrikate repräsentiere. In jedem Teilsystem $i \in I$ sei die Menge von K_i^r Anlagen zur Herstellung der Zwischenprodukte r-ter Stufe vorhanden. Vereinfacht wird hier $K_i^r \neq \emptyset$ für alle $i \in I$ und $r \in R$ angenommen.

Zwischen allen Teilsystemen werden Transportbeziehungen vorausgesetzt, so daß fehlende Fabrikate eines Typs von einem anderen Betriebsteil bezogen werden können. Die vertikale Verflechtung ist dann wie in Abb. 1 und die horizontale Verflechtung wie in Abb. 2 zu verstehen.

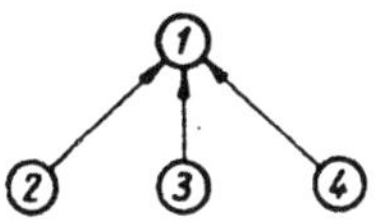

Abb. 1. $R = \{1, 2, 3, 4\}$

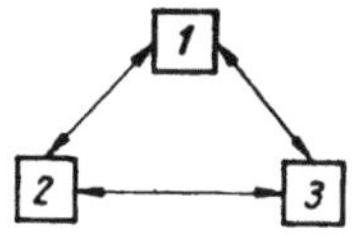

Abb. 2. $I = \{1, 2, 3\}$

[1] Technische Hochschule Karl-Marx-Stadt, Sektion Mathematik, DDR – 90 Karl-Marx-Stadt, Reichenhainer Str. 41.

Weiter nehmen wir an, daß eine eineindeutige Zuordnung der Zwischenprodukte jeder Stufe r zu den Endprodukten vorliegt, die es uns erlaubt, in jeder Stufe von einer Menge J von Produkten bzw. Halbfabrikaten zu sprechen.

Unter diesen Voraussetzungen steht die Aufgabe, einen solchen Plan zu bestimmen, der angibt, auf welcher Anlage jeder Stufe jedes Teilsystems welches Erzeugnis gefertigt und welcher Transport zwischen den einzelnen Betriebsteilen realisiert werden soll. Dieser Plan wird durch die Variablen

$$x^r_{ijk} \geqq 0, \quad j \in J, \quad k \in K^r_i, \quad i \in I, \quad r \in R, \tag{1}$$

und

$$z^r_{isj} \geqq 0, \quad i, s \in I, \quad j \in J, \quad r \in R - \{1\}, \tag{2}$$

festgelegt, wobei durch x^r_{ijk} die Menge der auf einer k-ten Anlage des i-ten Teilsystems gefertigten j-ten Produkte r-ter Stufe und z^r_{isj} die Menge der von $i \in I$ nach $s \in I$ $(s \neq i)$ zu transportierenden j-ten Produkte r-ter Stufe angeben.

Die Produktion und der Transport der Erzeugnisse werden innerhalb eines Planungszeitraums untersucht. Für diesen Zeitraum seien durch B^-_j, B^+_j untere und obere Schranken des Bedarfs an j-ten Endprodukten $(r = \{1\})$ vorgegeben, denen die Produktion gerecht werden muß. Diese Forderung führt auf die Restriktionen

$$B^-_j \leqq \sum_{i \in I} \sum_{k \in K^1_i} x^1_{ijk} \leqq B^+_j, \quad j \in J. \tag{3}$$

Für die Komplettierung der Produktion in der Stufe $r = \{1\}$ sind in den Teilsystemen $i \in I$ die Voraussetzungen durch die Eigenproduktion oder durch den Transport der Halbfabrikate nach i zu schaffen, d. h., es soll

$$\sum_{k \in K^1_i} x^1_{ijk} - \sum_{k \in K^r_i} x^r_{ijk} + \sum_{s \in I} (z^r_{isj} - z^r_{sij}) = 0, \quad j \in J, \quad i \in I, \tag{4}$$

$r \in R - \{1\}$, gelten.

Für jede Anlage $k \in K^r_i$ ist eine Kapazität $P^r_{ik} \geqq 0$ für den Planungszeitraum vorgegeben, die nicht überschritten werden darf. Mit $a^r_{ijk} \geqq 0$ sind Leistungskoeffizienten bezeichnet, die die benötigte Kapazität für die Herstellung einer j-ten Einheit eines Produktes r-ter Stufe auf der k-ten Anlage des i-ten Teilsystems bestimmen. Es wird

$$\sum_{j \in J} a^r_{ijk} x^r_{ijk} \leqq P^r_{ik}, \quad k \in K^r_i, \quad i \in I, \quad r \in R, \tag{5}$$

gefordert. Aus ökonomischen Erwägungen heraus ist der Betrieb daran interessiert, daß im Falle der Herstellung eines j-ten Produktes auf einer Anlage eine Mindestmenge hergestellt wird. Diese sei mit $\bar{x}^r_{ijk}$ bezeichnet, wobei

$$\bar{x}^r_{ijk} \leqq \min \{P^r_{ik}/a^r_{ijk}, B^-_j\} \quad \text{bei} \quad B^-_j > 0$$

und (6.1)

$$\bar{x}^r_{ijk} \leqq \min \{P^r_{ik}/a^r_{ijk}, B^+_j\} \quad \text{bei} \quad B^-_j = 0$$

gelte, $k \in K^r_i$, $r \in R$, $i \in I$, $j \in J$.

Die ökonomischen Forderungen beschreiben wir dann mit Hilfe der Ungleichungen

$$B_j^+ \operatorname{sign} x_{ijk}^r \geqq x_{ijk}^r \geqq \bar{x}_{ijk}^r \operatorname{sign} x_{ijk}^r, \quad k \in K_i^r, \quad i \in I, \quad r \in R, \quad j \in J . \tag{6}$$

Als Zielfunktion wird

$$\sum_{r \in R} \sum_{i \in I} \sum_{j \in J} \Big(\sum_{k \in K_i} W_{ijk}^r x_{ijk}^r + \sum_{s \in I} v_{isj}^r z_{isj}^r \Big) \rightarrow \min \tag{7}$$

betrachtet. Die Parameter W_{ijk}^r und v_{isj}^r sind dabei als Kosten oder Wertigkeiten zu interpretieren. Die Transportbewertungen seien für alle Indizes als positiv vorausgesetzt.

Bemerkung 1: Gegenüber der Aufgabe (1)–(7) treten in der echt realisierten Variante folgende Besonderheiten auf:

1) Nicht für alle $r \in R$ und $i \in I$ ist $K_i^r \neq \emptyset$.

2) Es ist in einigen Fällen keine eineindeutige Zuordnung der Halbfabrikate zu den Endprodukten möglich.

3) Nicht alle Produkte j der verschiedenen Stufen können auf allen Anlagen der entsprechenden Stufe gefertigt werden.

4) In den Bilanzbeziehungen (4) kann in der rechten Seite Null durch Parameter ϑ_{ij}^r ersetzt werden, die als geplanter Verkauf (bei $\vartheta_{ij}^r < 0$) oder Einkauf (bei $\vartheta_{ij}^r > 0$) von Erzeugnissen der r-ten Stufe angesehen werden können.

5) Zusätzlich werden Materialbeschränkungen berücksichtigt.

Diese Abweichungen vom Grundmodell (1)–(7) treten wohl äußerlich in Erscheinung, ändern jedoch nichts am Wesen der Aufgabe. Deshalb behandeln wir im weiteren nur die Aufgabe (1)–(7).

2. Analyse des mathematischen Problems (1)–(7)

Wir verwenden im weiteren die folgenden Bezeichnungen:

$$x = \{x_j\}_{j \in J}, \quad z = \{z_j\}_{j \in J},$$

$$x_j = \{x_{ijk}^r\}_{k \in K_i^r, i \in I, r \in R}, \quad z_j = \{z_{isj}^r\}_{i,s \in I, r \in R}, \quad j \in J,$$

$$A_j x_j = \{a_{ijk}^r x_{ijk}^r\}_{k \in K_i^r, i \in I, r \in R}, \quad P = \{P_{ik}^r\}_{k \in K_i^r, i \in I, r \in R},$$

$$f_j(x_j, z_j) = \sum_{r \in R} \sum_{i \in I} \Big(\sum_{k \in K_i^r} W_{ijk}^r x_{ijk}^r + \sum_{s \in I} v_{isj}^r z_{isj}^r \Big), \quad j \in J,$$

$$f(x, z) = \sum_{j \in J} f_j(x_j, z_j), \quad \text{Zielfunktion der Aufgabe (1)–(7)},$$

$$S = \{[x, z] : (1)\text{–}(6)\}, \quad \text{zulässiger Bereich der Aufgabe}$$

$$S_3^j = \{[x_j, z_j] : (1)\text{–}(4)\}, \quad j \in J, \quad S_3 = \prod_{j \in J} S_3^j,$$

$$S_1^j = \{[x_j, z_j] \in S_3^j : (6)\}, \quad j \in J, \quad S_1 = \prod_{j \in J} S_1^j$$

$$S_2 = \{[x, z] : (5)\} .$$

Damit ist die Aufgabe (1)–(7) in der Form

$$f(S) = \min \{f(x, z) : [x, z] \in S\} \tag{8}$$
$$= \min \{\sum_{j \in J} f_j(x_j, z_j) : [x_j, z_j] \in S_1^j, j \in J, [x, z] \in S_2\}$$

darstellbar. Sie ist eine Optimierungsaufgabe mit linearer Zielfunktion und einem nichtzusammenhängenden Bereich, die eine Überführung in eine gemischt-ganzzahlige lineare Optimierungsaufgabe ermöglicht. In ihrer allgemeinen Form werden Aufgaben vom Typ (1)–(7) in [3] untersucht. Wir bleiben bei der Formulierung der Aufgaben (1)–(7) oder (8) und zeigen, wie durch die Lösung der zu (1)–(7) bzw. (8) zweifach dualen Aufgabe eine Näherungslösung erhalten werden kann, die bei $|J| \gg \sum_{i \in I} \sum_{r \in R} |K_i^r|$ praktischen Anforderungen hinreichend gut gerecht wird.

Die Mengen S_3 und S_3^j sind konvex und polyedrisch, $j \in J$. Wir bezeichnen mit $\tilde{X}$ die Menge der Extremalpunkte einer Menge X. Dann gilt der

Satz 1: 1) *Jede Aufgabe vom Typ*

$$f_j(S_3^j) = \min \{f_j(x_j, z_j) : [x_j, z_j] \in S_3^j\}, \quad j \in J,$$

ist lösbar und besitzt eine optimale Lösung mit $[\hat{x}_j, \hat{z}_j] \in \tilde{S}_3^j, j \in J$. *Für alle optimalen Lösungen gilt* $[\hat{x}_j, \hat{z}_j] \in \operatorname{conv} \tilde{S}_3^j, j \in J$.

2) $\tilde{S}_3^j \doteq \{[x_j, z_j] \in S_3^j : x_{ijk}^r = B_j^-$ oder B_j^+ für genau ein Paar $(i, k) \in I \times K_i^r, i \in I$, und jedes $r \in R$, und sonst gleich Null, $z_{isj}^r = B_j^-$ oder B_j^+ für höchstens ein Paar $(i, s) \in I \times I$ und jedes $r \in R$ und sonst gleich Null$\}$, $j \in J$.

3) $\tilde{S}_3^j \subset S_1^j \subset S_3^j, j \in J$.

4) *Jede optimale Lösung von*

$$f_j(\tilde{S}_3^j) = \min \{f_j(x_j, z_j) : [x_j, z_j] \in S_3^j\}, \quad j \in J$$

löst die Aufgabe $f_j(S_1^j) = \min \{f_j(x_j, z_j) : [x_j, z_j] \in S_1^j\}, j \in J$.

Beweis: 1) Wir haben eine lineare Optimierungsaufgabe mit einer auf S_3^j nach unten beschränkten Zielfunktion. Dann ist die Existenz einer optimalen Extremalpunktlösung gesichert. Da die Koeffizienten des Transportaufwandes positiv sind, ist keine unbeschränkte Kante mit optimalen Lösungen möglich, und alle optimalen Lösungen sind als konvexe Linearkombination der Extremalpunkte darzustellen.

2) Diese Aussage gilt offensichtlich.

3) Wegen (6.1) und der vorangegangenen Aussage gilt bei $[x_j, z_j] \in \tilde{S}_3^j$ immer $x_{ijk}^r = 0$ oder $x_{ijk}^r \geqq \bar{x}_{ijk}^r$, $k \in K_i^r$, $i \in I$, $r \in R$, d. h. $[x_j, z_j] \in S_1^j$. Die Inklusion $S_1^j \subset S_3^j$ gilt trivialerweise, $j \in J$.

4) Auf Grund der Aussagen 1 und 3 des Satzes haben wir $f_j(\tilde{S}_3^j) = f_j(S_1^j) = f_j(S_3^j)$ und $\arg f_j(\tilde{S}_3^j) \subset \arg f_j(S_1^j), j \in J$, damit ist der Satz bewiesen. ∎

Wir bilden die Lagrangefunktion

$$L(x, z, u) = \sum_{j \in J} (f_j(x_j, z_j) + uA_j x_j) - uP$$

und die Aufgabe $L(u) = \min \{L(x, z, u) : [x, z] \in S_1\}$.

Bemerkung 2: 1) Nach dem Satz 1 gilt

$$L(u)=\min\{L(x,z,u):[x,z]\in S_1\}=\min\{L(x,z,u):[x,z]\in\tilde{S}_3\}.$$

2) $L(u)$ ist eine stückweise affin-lineare konkave Funktion auf $u\geqq 0$.

Wir bezeichnen mit

$$L=\max\{L(u):u\geqq 0\} \tag{9}$$

die *duale Aufgabe*. Es gilt dann der etwas modifizierte

Satz 2 ([4, 5]): *Bei $S\neq\emptyset$ sind die Aufgaben* (8) *und* (9) *lösbar, und es gilt die Ungleichung $L\leqq f(S)$.*

Die Lösung der dualen Aufgabe (9) führt auf eine großdimensionierte lineare Optimierungsaufgabe

$$\begin{aligned}L&=\max\{L(u):u\geqq 0\}=\max\{\min[L(x,z,u):[x,z]\in\tilde{S}_3]:u\geqq 0\}\\&=\max\{\sum_{j\in J}\min[f_j(x_j,z_j)+uA_jx_j:[x_j,z_j]\in\tilde{S}_3^j]-uP:u\geqq 0\}\\&=\max\{\sum_{j\in J}s_j-uP:s_j\leqq f_j(x_j^l,z_j^l)+uA_jx_j^l,\ l\in E_j,\ j\in J,\ u\geqq 0\},\end{aligned}$$

wobei

$$\{[x_j^l,z_j^l]:l\in E_j\}=\tilde{S}_3^j,\quad j\in J.$$

Die erneute Dualisierung ergibt dann

$$\begin{aligned}L=\min\{&\sum_{j\in J}\sum_{l\in E_j}f_j(x_j^l,z_j^l)\,v_{lj}:\sum_{j\in J}\sum_{l\in E_j}A_jx_j^lv_{lj}\leqq P,\\&\sum_{l\in E_j}v_{lj}=1,\ v_{lj}\geqq 0,\ l\in E_j,\ j\in J\}.\end{aligned} \tag{10}$$

Bemerkung 3: 1) Die Aufgabe (10) ist ein lineares Optimierungsproblem, das mit Spaltengenerierungsverfahren (vgl. [1, 5]) effektiv gelöst werden kann.

2) Mit $[x',z']=\{[x_j',z_j']\}_{j\in J}=\{\sum_{l\in E_j}v_{lj}[x_j^l,z_j^l]\}_{j\in J}$ erhalten wir aus (10) zulässige Lösungen von (8), wenn $v_{l_j,j}=1$, $v_{lj}=0$, $l\neq l_j$, $j\in J$. Die Aufnahme von Ganzzahligkeitsforderungen an die Variablen v_{lj} in der Aufgabe (10) führt auf eine zu (8) äquivalente Aufgabe.

3) Die Menge der Lösungen $\{[x_j',z_j']\}_{j\in J}$ für alle zulässigen $\{v_{lj}\}_{l\in E_j,j\in J}$ ist gleich der konvexen Hülle von S.

4) Wir haben in der Aufgabe (10) genau $|J|+\sum_{i\in I}\sum_{r\in R}|K_i^r|$ Nebenbedingungen.

In Anlehnung an [6] gilt der

Satz 3: *Eine beliebige Basislösung der Aufgabe* (10) *hat wenigstens* $\max(0,|J|-\sum_{i\in I}\sum_{r\in R}|K_i^r|)$ *Indizes $j\in J$ mit $v_{l_j,j}=1$, $v_{lj}=0$, $l\neq l_j$.*

Beweis: Eine beliebige zulässige Basislösung besitzt nicht mehr als $|J|+\sum_{i\in I}\sum_{r\in R}|K_i^r|$ positive Komponenten. Da in jeder Gleichung wenigstens eine Komponente positiv sein muß, können höchstens in $\sum_{i\in I}\sum_{r\in R}|K_i^r|$ Gleichungen mehr als eine Komponente von v_{lj} positiv sein, was den Satz beweist. ∎

Bemerkung 4: 1) Bei

$$|J| \gg \sum_{i \in I} \sum_{r \in R} |K_i^r| \tag{11}$$

erhält man für die überwältigende Mehrheit der Produkte die Zulässigkeit bezüglich der Forderung (6).

2) Wegen der Bemerkung 3.3 genügen alle Lösungen $[x', z']$ den Restriktionen (1)–(5).

3) Die Ungleichung $L \leqq f(S)$ erlaubt die Schlußfolgerung, daß eine aus einer optimalen Lösung von (10) abgeleitete Lösung $[x', z']$, die entsprechend korrigiert wird, einen Funktionswert mit nur geringer Abweichung vom Optimum besitzt.

4) Diese Vorgehensweise erscheint prinzipiell als ein effektiver Zugang zur Lösung großer strukturierter diskreter Optimierungsaufgaben.

5) Die Ungleichung (11) kann für viele praktische Situationen als wahr angenommen werden. Daraus ergibt sich die Schlußfolgerung, daß einer Zerlegung der Aufgaben nach den Produktionsstufen oder der territorialen Verflechtung häufig die Zerlegung nach Produkten vorzuziehen ist.

3. Rechentechnische Ergebnisse und ökonomische Auswertung

Die Lösung der Aufgabe (10) mit einem Spaltengenerierungsverfahren führt zu einer Variante des Verfahrens von Dantzig/Wolfe [1, 5], in der die optimalen Lösungen der Teilaufgaben vom Typ $f_j(S_3)$, $j \in J$, leicht bestimmt werden können. Die Aufgabe (10) kann als Zentralaufgabe des Verfahrens von Dantzig/Wolfe für die äquivalente lineare Optimierungsaufgabe (1)–(5), (7) angesehen werden. Obwohl das Spaltengenerierungsverfahren durchaus effektiver sein kann als eine Standardmethode, wird es gegenwärtig nicht zur Lösung des vorliegenden praktischen Problems herangezogen, da keine anwendungsbereiten Programme vorliegen. Deshalb wird zur Lösung des konkreten Modells (1)–(5), (7) für den VEB Aerosol-Automat Karl-Marx-Stadt die Simplexmethode, wie sie im VOPS OPSI DOS/ES bzw. PS OPSI OS/ES vom VEB Robotron realisiert wurde, benutzt. Die Dimension der linearen Optimierungsaufgabe (1)–(5), (7) – 500 bis 1200 Zeilen – erfordert nicht unmittelbar die Anwendung von Zerlegungsmethoden. Das VOPS OPSI wird für dieses Problem den Ansprüchen an Rechenzeit und Genauigkeit in jeder Weise gerecht. Mit der Anwendung des PS OPSI werden Beschränkungen für die Zeilenzahl des Modells wirksam. Es bezeichne

n_1 – Anzahl der Fertigerzeugnisse
n_2 – Anzahl aller Anlagen
n_3 – Anzahl aller Teilsysteme
n_4 – Anzahl aller Halbfabrikate ($n_1 + n_4 = |J|$)
n_5 – Anzahl der zu bilanzierenden Rohstoffe.

Dann muß gelten $n_1 + n_2 + n_3 \cdot n_4 + n_5 \leqq 4094$, wenn nur eine Zielfunktion vorgegeben ist. Im vorliegenden Problem werden dabei folgende Größen erreicht:

$$n_1 = 140 \ldots 350$$
$$n_2 = 73 \ldots 150$$
$$n_3 = 4 \ldots 5$$
$$n_4 = 98 \ldots 355$$
$$n_5 = 260 \ldots 400 .$$

Der Besetzungsgrad der Problemmatrix beträgt 0,74 %. Die Eingabe der Problemdaten erfolgt im LK-Format über Lochkarte (LK), Magnetband (MB) oder Magnetplatte (MP). Es ist möglich, die Problemdaten auf externen Speichermedien abzuspeichern und diese für weitere Rechnungen durch – oft nur geringfügige – Korrekturen auf den aktuellen Stand zu bringen. Bei Anwendung des PS OPSI kann diese Speicherung in Form von EDO-Dateien erfolgen, die auch für andere, vom VEB Robotron gelieferte Programmpakete oder -systeme verwertbar sind. Die Ausgabe der optimalen Lösung kann im VOPS OPSI bzw. im PS OPSI über Drucker in der Standardform oder im TRANSFER-Format auf externe Speichermedien erfolgen, speziell im PS OPSI auch in EDO-Dateien. In den meisten praktischen Anwendungsfällen, so auch im vorliegenden, genügen diese Möglichkeiten jedoch nicht allen betrieblichen Wünschen. Im OPSI sind einmal die Namen durch 8 Zeilen beschränkt, zum anderen wären oft auch Nachfolgeoperationen wünschenswert, um die Drucklisten anwendungsbereiter zu machen. Der vom PP EDO gelieferte Komfort hierzu ist noch immer zu gering. Deshalb erstellten wir ein betriebsspezifisches Druckprogramm (PL/I), das die optimale Lösung der linearen Optimierungsaufgabe, wie sie im TRANSFER-Format auf externen Speichermedien vorliegt, nach betrieblichen Gesichtspunkten auswertet und zugleich betriebliche Bezeichnungen und Erzeugnisnummern ausdruckt. Ein solches Programm kann im VOPS OPSI als Nachfolgeprogramm, im PS OPSI als Anwenderprogramm eingesetzt werden.

Die Verwendung eigener Drucklisten erspart einen wesentlichen Teil der Auswertung des Standarddruckes. Die Verwaltungsbereiche (Materialplanung, Produktionsplanung, Leiter der Teilsysteme u. a.) erhalten sofort verwendbare, für sie (auch ohne Kenntnis des Modells und des Programms) auswertbare Unterlagen zu ihrem Arbeitsbereich.

Die bei der vorliegenden praktischen Aufgabenstellung errechneten suboptimalen Lösungen bestätigen die Bemerkung 4.1. Nicht mehr als drei von etwa 240 Variablen x_{ijk}^r weisen die dort angeführte Eigenschaft nicht auf ($|J| \approx 240$). Für n_2 gilt $\sum_i \sum_r |K_i^r| \approx 70$, wobei aber nur die Kapazität von 12 Anlagen als echte Beschränkung auftritt, d. h., die Voraussetzung (11) ist hier erfüllt (vgl. Tab. 1).

Das beschriebene Modell dient zur Erstellung von Jahresplänen. Bisherige Testrechnungen ergaben eine Produktionssteigerung von etwa 2,8 % bezüglich der Stückzahl gegenüber der im gleichen Zeitraum realisierten Produktion. Gleichzeitig ist es möglich geworden, auf verschiedenen Anlagen Reserven aufzudecken, da die Auslastung der Anlagen errechnet und gedruckt wird.

Tabelle 1

Nummer des Beispiels	Zeilenzahl	Spaltenzahl	Besetzungsgrad der Matrix (%)	Rechenzeit (min) ES 1040	Anzahl der Verletzungen von (6)
1	321	487	0,79	8.51	1
2	325	501	0,75	8.39	2
3	568	544	0,73	7.73	2
4	577	543	0,74	7.83	1
5	577	543	0,75	14.08	2
6	577	543	0,74	8.01	1
7	579	543	0,75	8.02	3

Die dargestellten Untersuchungen wurden im Rahmen eines Wirtschaftsvertrages der Technischen Hochschule Karl-Marx-Stadt mit dem VEB Aerosol-Automat Karl-Marx-Stadt von Mitarbeitern und Studenten der Sektion Mathematik durchgeführt.

Literatur

[1] Beer, K.: Lösung großer linearer Optimierungsaufgaben. VEB Deutscher Verlag der Wiss., Berlin 1977.
[2] Beer, K., J. Käschel und K. Richter: Zur Lösung einer großen strukturierten diskreten Optimierungsaufgabe mit VOPS OPSI. Eingereicht in Rechentechnik/Datenverarbeitung.
[3] Беркович, М. М., и А. А. Животовский: Решение одной задачи размещения. Изв. АН СССР, Техн. киберн. (1969) 3.
[4] Geoffrion, A. M.: Lagrangian relaxation for integer programming. Math. Program. Study 2 (1974), 82–114.
[5] Lasdon, L. S.: Optimization theory for large systems. London, MacMillan, 1970.
[6] Manne, A. S.: Programming of economic lot sizes. Man. Science 4 (1958) 2, 115–135.
[7] Anwenderhandbuch 1–3, VOPS OPSI/DOS/ES, VEB Kombinat Robotron.

Резюме

Исследуется дискретная модель составления оптимального плана производства для нескольких продуктов, производственных отделений, производственных линий и ступеней производства. Если число продуктов значительно больше числа линий по всем отделений и ступеней, то соответствующая дважды двойственная задача линейного программирования дает хорошое приближенное решение исходной задачи. Приводятся некоторые вычислительные и экономические результаты применения модели.

Summary

A model for developing an optimal production plan of several products, workshops, plants and stages of production under certain discreteness conditions is investigated. If the number of products is essentially greater than the number of plants in all workshops and stages of production, an appropriate twice-dual linear programming problem yields a feasible solution of the original problem. Some computational and economic results obtained from the application of the model are presented.

Eingereicht bei der Redaktion: 17. 2. 1978; revidierte Fassung: 5. 6. 1978

Math. Operationsforsch. Statist., Ser. Optimization, Vol. 10 (1979) No. 1, 93–96

One - Processor Scheduling with Nonuniform Ready Times[1]

MILAN VLACH[2]

Summary: This paper is concerned with one-machine scheduling problems allowing non-simultaneous jobs arrivals. For problems in which all jobs are simultaneously available HELD and KARP developed a dynamic programming algorithm for finding an ordering that minimizes the total sum of arbitrary cost functions. This paper extends the approach of HELD and KARP to the problems with nonsimultaneous arrivals and arbitrary non-decreasing cost functions. The algorithm is illustrated by a problem of minimizing the number of tardy jobs.

Suppose we have n jobs to be processed on a continuously available processor that can handle only one job at a time. Each job j ($j=1, 2, \ldots, n$) has an individual ready time r_j, before which its execution cannot begin. Job j takes t_j time units to be processed. With job j a nondecreasing real function f_j is associated, whose value $f_j(t)$ represents the costs incurred if job j is completed at time t. Any permutation p of the set $N=\{1, 2, \ldots, n\}$ leads to a completion time $u_j(p)$ for job j by processing the jobs in order of their occurrence on the list $\langle p(1), p(2), \ldots, p(n)\rangle$ and starting each job at the earliest possible time. The objective is to find a permutation that minimizes the function

$$f(p) := \sum_{j=1}^{n} f_j(u_j(p)) .$$

In recent years considerable attention has been given to the problems with

$$f_j(t) := k_j \max(0, t-d_j)$$

and

$$f_j(t) := k_j \max(0, \operatorname{sign}(t-d_j))$$

where d_j is a given due date for job j and k_j is a given positive number (often $k_j=1$ for each j or $k_j=\frac{1}{n}$ for each j). Most results offered to date have required the simplifying assumption

$$r_1 = r_2 = \ldots = r_n .$$

[1] The paper was presented at the conference „Mathematische Optimierung – Theorie und Anwendungen" in Eisenach (GDR), December 5 to 10, 1977.

[2] MFF UK, Malostranske nám 25, ČS – 11800 Praha 1, ČSSR.

Even in this situation the mentioned problems are NP-complete and it is therefore not surprising that exact methods for solving the general problem involve some kind of enumeration by dynamic programming or branch-and-bound-technique.

Dynamic programming was applied to sequencing problems already in 1962 by HELD and KARP [1]. Improved algorithms for special cases were found by LAWLER and MOORE [2]. In this paper a dynamic programming algorithm for solving the problems with general nondecreasing functions $f_1, f_2, \ldots, f_n$ and with not necessarily identical ready times is given. To illustrate the algorithm a problem of minimizing the number of tardy jobs is solved in detail.

Let S be a nonempty subset of the set N and let $F(S)$ denote the optimal value of the objective function for the problem whose set of jobs is S. The value $F(S)$ may be achieved for several permutations of the set S and to each of these permutations corresponds certain total completion time. Let $T(S)$ denote the smallest of these times and let p_S be one of the permutations which give $T(S)$. Defining

$$F(\emptyset)=0, \quad T(\emptyset)=0 \tag{1}$$

the principle of optimality of dynamic programming gives

$$F(S)=\min_{k\in S} [F(S\setminus\{k\})+f_k\,(t_k+\max\,(r_k, T(S\setminus\{k\})))] \tag{2}$$

for each nonempty subset S of N.

The recursion relation (2) indicates that in order to calculate the value of F for any particular set S of s jobs one first must know the value of F for s subsets of $s-1$ jobs. Therefore by using (1), (2), and the input data t_k, r_k, f_k one can first calculate

$$F(\{j\})=f_j\,(t_j+r_j), \quad j=1, 2, \ldots, n\,,$$

then the value of F for all two-element subsets and so on. At the final stage the procedure determines $F(N)$. Keeping track of where minima occurred at each stage one will be able to construct the optimal permutations.

To illustrate the method, consider the problem with

$$f_j(t)=\max\,(0, \operatorname{sign}\,(t-d_j))$$

and with the following job set

j	1	2	3	4
r_j	0	2	4	1
t_j	3	5	3	8
d_j	7	9	15	15

The results of dynamic programming calculations for proper subsets of $N=\{1, 2, 3, 4\}$ are displayed in the following tables:

S	$F(S)$	$T(S)$	p_S
{1}	0	3	⟨1⟩
{2}	0	7	⟨2⟩
{3}	0	7	⟨3⟩
{4}	0	9	⟨4⟩

S	$F(S)$	$T(S)$	p_S
{1, 2}	0	8	⟨1, 2⟩
{1, 3}	0	7	⟨1, 3⟩
{1, 4}	0	11	⟨1, 4⟩
{2, 3}	0	10	⟨2, 3⟩
{2, 4}	1	14	⟨4, 2⟩
{3, 4}	0	12	⟨4, 3⟩

S	$F(S)$	$T(S)$	p_S
{1, 2, 3}	0	11	⟨1, 2, 3⟩
{1, 2, 4}	1	16	⟨1, 2, 4⟩, ⟨1, 4, 2⟩
{1, 3, 4}	0	14	⟨1, 4, 3⟩
{2, 3, 4}	1	17	⟨4, 3, 2⟩

As example, consider the sets $S=\{1, 2, 4\}$ and $S=\{2, 3, 4\}$. In the first case one has

$$F(\{1, 2\})+f_4(t_4+\max(r_4, T(\{1, 2\})))=f_4(8+\max(1, 8))=1\ ,$$
$$F(\{1, 4\})+f_2(t_2+\max(r_2, T(\{1, 4\})))=f_2(5+\max(2, 11))=1\ ,$$
$$F(\{2, 4\})+f_1(t_1+\max(r_1, T(\{2, 4\})))=1+f_1(3+\max(0, 14))=2\ .$$

Only two permutations from six possible give the value $F(\{1, 2, 4\})$, namely $\langle 1, 2, 4\rangle$ and $\langle 1, 4, 2\rangle$. For both of them the total completion time is 16. Therefore $T(\{1, 2, 4\})=16$ and both the permutations may be taken as $p_{\{1,2,4\}}$. In the second case one has

$$F(\{2, 3\})+f_4(t_4+\max(r_4, T(\{2, 3\})))=f_4(8+\max(1, 10))=1\ ,$$
$$F(\{2, 4\})+f_3(t_3+\max(r_3, T(\{2, 4\})))=1+f_3(3+\max(4, 14))=2\ ,$$
$$F(\{3, 4\})+f_2(t_2+\max(r_2, T(\{3, 4\})))=f_2(5+\max(2, 12))=1\ .$$

Again the value $F(\{2, 3, 4\})$ is achieved for two permutations $\langle 2, 3, 4\rangle$ and $\langle 4, 3, 2\rangle$. However the total completion time corresponding to the permutation $\langle 2, 3, 4\rangle$ is 18 and to the permutation $\langle 4, 3, 2\rangle$ is 17. Therefore $T(\{2, 3, 4\})=17$ and $p_{\{2,3,4\}}=$ $=\langle 4, 3, 2\rangle$.

Finally for the set $N=\{1, 2, 3, 4\}$ one has

$$F(\{1, 2, 3\})+f_4(t_4+\max(r_4, T(\{1, 2, 3\})))=1\ ,$$
$$F(\{1, 2, 4\})+f_3(t_3+\max(r_3, T(\{1, 2, 4\})))=2\ ,$$
$$F(\{1, 3, 4\})+f_2(t_2+\max(r_2, T(\{1, 3, 4\})))=1\ ,$$
$$F(\{2, 3, 4\})+f_1(t_1+\max(r_1, T(\{2, 3, 4\})))=2\ .$$

Therefore $F(N)=1$ and this value is achieved for the permutations $\langle 1, 2, 3, 4\rangle$ and $\langle 1, 4, 3, 2\rangle$. Both of them are optimal. In the first case job 4 is tardy, in the second case job 2 is tardy.

References

[1] Held, M., and R. Karp: A Dynamic Programming Approach to Sequencing Problems. SIAM Journal on Applied Mathematics **10** (1962), 196–210.
[2] Lawler, E. L., and J. M. Moore: A Functional Equation and Application to Resource Allocation and Sequencing Problems. Management Science **16** (1969), 77–84.

Zusammenfassung

Es wird das Reihenfolgeproblem der Bearbeitung von nicht gleichzeitig eintreffenden Aufträgen für eine Maschine betrachtet. Für Probleme, bei denen alle Aufträge gleichzeitig eintreffen, wurde von Held und Karp ein Algorithmus mittels dynamischer Optimierung zur Bestimmung einer Reihenfolge, die die Summe von beliebigen Kostenfunktionen minimiert, angegeben. In dieser Note wird die Methode von Held und Karp auf Probleme mit nicht gleichzeitig eintreffenden Aufträgen und monoton nichtfallenden Kostenfunktionen erweitert. Der Algorithmus wird anhand der Minimierung der Anzahl von nicht erfüllten Aufträgen erläutert.

Резюме

Рассматривается проблема упорядочения работ на одной машине в ситуации допускающей неодновременное поступление работ. Для задач, в которых работы поступают одновременно, Хелд и Карп разработали вычислительную схему динамического программирования, которая применима в случае суммы произвольных функций. В статье расширяется эта схема на ситуации, в которых работы могут поступать неодновременно и минимизируется сумма произвольных неубывающих функций. Алгорифм иллюстрируется решением задачи минимизации числа невыполненых сроков.

Eingereicht bei der Redaktion: 7. 12. 1977

Math. Operationsforsch. Statist., Ser. Optimization, Vol. 10 (1979) No. 1, 97–100

On Pontryagin Maximum Principle for Systems with Non-one-point Target Sets and Systems with Additional Constraints[1]

S. Rolewicz[2]

Summary: In the paper a linear system $(X \xrightarrow[C]{} Y)$ is considered. We are looking for $\inf\{F(x): x \in U,\ Cx = y_0\}$, where F is a convex functional and U is a closed convex set. Sufficient conditions warrantying that the Maximum Principle of Pontryagin implies closedness of CX are given.

Let a linear system (see [2]) be given

$$(X \xrightarrow[C]{} Y). \tag{1}$$

Let $F(x)$ be a continuous convex functional defined on X. We say that Pontryagin Maximum Principle (briefly PMP) holds if for all $y_0 \in CX$ there is a continuous linear functional $\varphi_0 \in Y^*$ such that

$$\inf\{F(x): Cx = y_0\} = \inf\{F(x): \varphi_0(Cx) = \varphi_0(y_0)\}. \tag{2}$$

By Hahn-Banach theorem we trivially obtain that if CX is closed and $\Gamma_r = C\{x: F(x) \leqq r\}$ are closed then PMP holds.

The following theorem is in a certain sense converse to this fact.

Theorem 1 [2]: *Let Γ_r be closed. If* PMP *holds then CX is closed.*

The main purpose of the paper is to discuss extensions of theorem 1 in two directions. The first direction will be the case when target set is not one-point set. The second one will be the case when additional constraints on x exist.

Let Ω be an arbitrary closed convex set in Y. We say that PMP holds if for all $y_0 \in CX$ there is $\varphi_0 \in Y^*$ such that

$$\inf\{F(x): Cx \in y_0 + \Omega\} = \inf\{F(x): \varphi_0(Cx) = \varphi_0(y_0) + \inf_{y \in \Omega} \varphi_0(y)\}. \tag{3}$$

Such problem was considered in [2], for Ω being a closed ball. It is easy to verify that if $\Gamma_r + \Omega$ are closed that PMP holds if and only if $(\Gamma_r + \Omega) \cap CX$ has interior in CX. It trivially implies that if either CX is closed or Ω has interior, that PMP holds.

It is not difficult to give an example of systems in which Γ_r do not have interiors in CX and Ω also do not have interior such that $\Gamma_r + \Omega$ have interiors.

[1] The paper was presented at the conference „Mathematische Optimierung – Theorie und Anwendungen" in Eisenach (GDR), December 5 to 10, 1977.

[2] Instytut Matematyczny Polskiej Akademii Nauk, 00950 Warszawa, ul. Sniadeckich 8, Skr. poczt. 137, Poland.

Example 1: Let $X=\mathfrak{L}^2\times\mathfrak{L}^2=Y$. Let T be a compact non-finite dimensional operator acting in $\mathfrak{L}^2$, $T\colon \mathfrak{L}^2\to\mathfrak{L}^2$. Let $F(x_1, x_2)=\sqrt{\|x_1\|^2+\|x_2\|^2}$ be a HILBERT norm in X. Let $C(x_1, x_2)=(Tx_1, x_2)$. Let $\Omega=(K, TK)$, where K denotes the closed unit ball in $\mathfrak{L}^2$. It is easy to verify that Γ_r and Ω do not have relative interior (i.e. interior in the smallest linear manifold containing Γ_r (or Ω)) and on the other hand $\Gamma_r+\Omega$ is a closed convex set with interior. Thus PMP holds.

The presented above example is strictly related to a following problem.

Problem 1: How ought be properties of a BANACH space X warranting that for two arbitrary closed convex sets A, B such that $A+B$ has interior there are two closed linear manifolds X_1, X_2 such that $X_1+X_2=X$, $A\cap X_1$ ($B\cap X_2$) has interior in X_1 (correspondingly in X_2)?

It follows from results of FILLMORE and WILLIAMS [1] that it is so if X is a HILBERT space. There is also a counterexample for $X=\mathfrak{L}^1$ (oral communication of P. WOJTASZCZYK).

Now we shall consider a problem with additional constraints.

Let U be a closed convex set in X. We say that PMP holds if for all $y_0\in CU$ there is $\varphi_0\in Y^*$

$$\inf\{F(x)\colon Cx=y_0,\ x\in U\}=\inf\{F(x)\colon \varphi_0(Cx)=\varphi_0(y_0),\ x\in U\}\,. \tag{4}$$

Now we recall notions of algebraic interior and algebraic boundary. Let K be a convex spaces in a linear space E. We say that p is an *algebraic internal point* if each prime line L passing through p either has with L only one common point p, $L\cap K=\{p\}$, or $L\cap K$ is an interval such that p belongs to its interior with respect to the prime line L. The set of all algebraic internal points is called *algebraic interior* of K and it is denoted $\mathrm{Int}_a K$. The set $K\backslash\mathrm{Int}_a K$ is called *algebraic boundary* of the set K and it is denoted $\mathrm{Fr}_a K$. The PMP without constraints given by formula (2) can be interpreted such that each $p\in\mathrm{Fr}_a\Gamma_r$, p is a point of support of Γ_r (it means there is a linear continuous function φ such that $\varphi(p)=\inf_{y\in\Gamma_r}\varphi(y)<\sup_{y\in\Gamma_r}\varphi(y)$) (see [2]). In the same way PMP can be interpreted such that for each $p\in\mathrm{Fr}_a C\,\{x\in U\colon F(x)\leq r\}\cap C\,\{x\in U\colon F(x)=r\}$ p is a point of support.

The effect that p belongs also to $C(U)$ gives us that PMP does not imply closedness of CX.

Example 2: Let $X=Y=\mathfrak{L}^2$. Let $C(\{x_n\})=\left\{\frac{x_n}{n}\right\}$. Let $F(x)=\max\,(\|x\|_{\mathfrak{L}^2}, 2\,|x_1|)$. Let $U=\left\{x=\{x_n\}\colon x_1\geq\left(\sum_{i=2}^{\infty}x_i^2\right)^{1/2}\ \|x\|\leq 4\right\}$. It is easy to verify that CX is not closed. The sets Γ_r are weakly compact, hence closed. The set CU is also weakly compact, hence closed. Observe that $\mathrm{Fr}_a\Gamma_r\cap C(U)\subset\{y\colon y_1=2^r\}$. Thus for all $p\in\mathrm{Fr}_a\Gamma_r\cap C(U)$ linear functional $\varphi(y)=y_1$ supports Γ_r at the point p. It implies that (4) holds.

Now we shall discuss, which additional restriction on $F(x)$ warrants, that PMP implies closedness of CX. Strict convexity is not sufficient as follows from

Example 3: Let X, Y, U, C be as in Example 2. Let

$$F(x)=\max\left(\sum_{i=1}^{\infty}\left(\frac{x_i}{i}\right)^2, \frac{1}{4}\|x\|_{\ell^2}\right).$$

Of course $F(x)$ is strictly convex. Observe that $\mathrm{Fr}_a\Gamma_r\cap C(U)\subset\{y: \|y\|=r\}$. Thus for all $p\in\mathrm{Fr}_a\Gamma_r\cap C(U)$ linear functional $\langle y, p\rangle$ supports Γ_r at the points p. It implies that (4) holds.

Theorem 2: *Let X be a reflexive* BANACH *space. Let $F(x)$ be uniformly convex* i.e. *for arbitrary $\varepsilon>0$, there is $\delta(\varepsilon)>0$ such that*

$$F\left(\frac{x_1+x_2}{2}\right)\leqq\frac{F(x_1)+F(x_2)}{2}-\delta \tag{5}$$

for all x_1, x_2 such that $\|x_1-x_2\|\geqq\varepsilon$. If the set $\mathrm{Fr}_a\Gamma_r\cap C\,\{x\in\mathrm{Int}\,U, F(x)\leqq r\}$ is not empty, then PMP *implies that CX is closed.*

The proof of the theorem is based on a following lemma

Lemma 3: *Let $F(x)$ be strictly convex. Let $K_{r_0}=\{x: F(x)\leqq r_0\}$. Let $F(x_0)=r_0$. Then there is a linear continuous functional f and c such that $f(x_0)>c$ and the diameter of the set $\{x\in K_{r_0}: f(x)\geqq c\}$ is less than ε.*

Proof: The sets K_r are closed. By definition $x_0\notin K_{r_0-\delta/2}$. Thus there is a linear continuous functional f and a number c such that

$$f(x_0)>c \tag{6}$$

and

$$f(z)\leqq c \quad \text{for} \quad z\in K_{r_0-\delta/2}\,. \tag{7}$$

The functional f and c satisfies conditions of the lemma. In fact, suppose that diam $(\{x\in K_{r_0}: f(x)\geqq c\})>\varepsilon$. Then there are x_1, x_2 such that $\|x_1-x_2\|\geqq\varepsilon$, $x_1, x_2\in\in K_{r_0}$, $f(x_1), f(x_2)\geqq c$. If implies that

$$\frac{F(x_1)+F(x_2)}{2}\leqq r$$

and

$$F\left(\frac{x_1+x_2}{2}\right)\leqq r-\delta \tag{8}$$

which is contradictory to (7).

Proof of the theorem: Under our hypothesis there is $p\in\mathrm{Fr}_a\Gamma_r\cap C\,\{x\in\mathrm{Int}\,U: F(x)\leqq r\}$. Let $x_0\in\mathrm{Int}\,U$ be an element such that $Cx=p$ and $F(x_0)=r$. Since $F(x_0)$ is strictly convex x_0 is uniquely determined. By lemma 3 there is a linear continuous functional f and a constant c such that $f(x_0)>c$ and

$$\{x: F(x)\leqq r, f(x)\geqq c\}\subset\mathrm{Int}\,U\,. \tag{9}$$

Let $V=\left\{x\in X: F(x)\leqq r, f(x)\geqq\frac{c+f(x_0)}{2}\right\}$. The set V is weakly compact. The set

$$V_1=\{x\in X: f(x)\leqq c, F(x)\leqq r\} \tag{10}$$

is also weakly compact. Hence the sets CV and CV_1 are weakly compact too. Thus there is a linear continuous functional φ separating CV and CV_1. Thus there is a constant c_1 such that for functional $f_1 = C^*\varphi$

$$f_1(x) > c_1 \quad \text{for} \quad x \in V$$

and

$$f_1(x) < c_1 \quad \text{for} \quad x \in V_1 .$$

Let

$$W = \{x \in X : F(x) \leqq r, f_1(x) \geqq c_1\} .$$

Of course

$$W \cap V_1 = \emptyset .$$

The set W has non empty interior and $W \subset \text{Int } U$. Each point p belonging to the algebraic boundary of $C(W)$ is a point of support. In fact if $p = Cx$, where $f(x) = c_1$, then φ is a functional of support. If p is not of the form described above then $p \in \text{Fr}_a \Gamma_r \cap C \{x \in U : F(x) = r\}$ and p is a point of support. Therefore by Lemma of WOJTASZCZYK [4] we obtain in similar way as in Theorem 1, that CX is closed.

Problem 2: Is the hypothesis that $\text{Fr}_a \Gamma_r \cap C \{x \in \text{Int } U : F(x) \leqq r\} \neq 0$ essential?

References

[1] FILLMORE, P., and J. WILLIAMS: On operator range. Advance in Matematics 7 (1971), 254–281.
[2] ROLEWICZ, S.: Funktionalanalysis und Steuerungstheorie. Springer Verlag 1976.
[3] ROLEWICZ, S.: On convex sets containing only points of support. Comm. Math. vol. spec. I, 279–281.
[4] WOJTASZCZYK, P.: A theorem on convex sets related to the Abstract PONTRYAGIN Maximum Principle. Bull. Acad. Pol. Sc. 21 (1973), 93–95.

Резюме

В работе рассматривается линейная система $(X \xrightarrow[C]{} Y)$. Мы ищем inf $\{F(x) : x \in U, Gx = y_0\}$, где F выпуклый функционал а U выпуклое замкнутое множество. Дается достаточные условия, при которых из Принципа Максимума ПОНТРЯГИНА следует, что CX – замкнутое множество.

Sommaire

Dans cette note, on considère un système linéaire $(X \xrightarrow[C]{} Y)$. On cherche inf $\{F(x) : x \in U, Cx = y_0\}$, où F est une fonctionelle convexe U est ensemble convexe fermé. Nous donnons des conditions suffisantes pour que le principe de maximum implique la fermature de CX.

Eingereicht bei der Redaktion: 16. 2. 1978

Math. Operationsforsch. Statist., Ser. Optimization, Vol. 10 (1979) No. 1, 101–110

A-priori-Abschätzungen von Optimalwerten zu Steuerungsproblemen. Teil I[1]

ROLF KLÖTZLER[2]

Zusammenfassung: Vorliegende Arbeit ist eine Anwendung eines neuen allgemeinen Dualitätsprinzips auf Variationsprobleme einfacher Integrale in Parameterdarstellung unter Zustandsbeschränkungen. Es werden eine zweiseitige Abschätzung des Optimalwertes und deren numerische Realisierung mittels graphentheoretischer Methoden abgeleitet.

1. Einleitung

In jüngster Zeit wurde vom Verfasser ein neues und sehr allgemeines Dualitätsprinzip zu stetigen Steuerungsproblemen angegeben, das ohne jegliche Konvexitätsvoraussetzungen gilt (vgl. [3]–[6]). Unter Beschränkung auf einfache Integrale und Festrandprobleme lautet es in einer für unsere Zwecke etwas vereinfachten Form wie folgt:

Gegeben sei das Steuerungsproblem

$$J(x, u) = \int_0^T r(t, x, u)\, dt \to \text{Min} \tag{1}$$

bez. aller auf $[0, T]$ definierten vektorwertigen
Zustandsfunktionen $x = (x^1, \ldots, x^n) \in X$ und
Steuerungsfunktionen $u = (u^1, \ldots, u^l) \in U(x)$ mit
$X := \{x \in W_p^{1,n}(0, T) \mid (t, x(t)) \in \bar{G}$ auf $[0, T]$; $x(0) = x_0, x(T) = x_T\}$,
$U(x) := \{u \in L_p^l(0, T) \mid u(t) \in V(t, x(t))$ fast überall auf $(0, T)\}$
unter den Nebenbedingungen (*Zustandsgleichungen*)

$$\dot{x}^i = g^i(t, x, u) \quad (i = 1, \ldots, n).$$

Dabei seien $p = \text{const} > 1$, G ein Normalgebiet des E^{n+1} im Sinne von [7], S. 14, und V eine mengenwertige Abbildung von $\bar{G}$ in den R^l. Die Funktionen r und g^i seien reell und stetig auf $\bar{G} \times E^l$, und die Menge $\mathfrak{P}$ aller zulässigen Paare $\langle x, u \rangle$ – kurz *Prozesse* genannt – sei nicht leer; außerdem sei $J(x, u)$ für alle $\langle x, u \rangle \in \mathfrak{P}$ erklärt.

Dann gilt für jeden Prozeß $\langle x, u \rangle$ und jede auf $\bar{G}$ stückweise stetig differenzier-

---

[1] Vortrag, gehalten auf der Tagung „Mathematische Optimierung – Theorie und Anwendungen" in Eisenach vom 5.–10. 12. 1977.

[2] Karl-Marx-Universität Leipzig, Sektion Mathematik, DDR-701 Leipzig, Karl-Marx-Platz.

bare Funktion S (d. h. $S\in D^1(\bar{G})$ im Sinne von [7], S. 145) die *Dualitätsbeziehung*

$$J(x, u) \geqq L(S) \tag{2}$$

mit $$L(S) := -\int_0^T \sup_{\xi\in Q(t)} \delta_S(t, \xi)\, dt + S(t, x(t))\Big|_0^T ,$$

wobei δ_S den *Defekt der* HAMILTON-JACOBIschen *Differentialgleichung* bezeichnet und durch

$$\delta_S(t, \xi) := S_t(t, \xi) + \sup_{v\in V(t,\xi)} H(t, \xi, v, S_\xi(t, \xi))$$

definiert ist. Außerdem sind

$$Q(t) := \{\xi\in E^n \mid (t, \xi)\in\bar{G}\} \quad \forall t\in(0, T)$$

und H die PONTRJAGINsche *Funktion* zu (1), d. h.,

$$H(t, \xi, v, y) := -r(t, \xi, v) + y_i g^i(t, \xi, v) ,$$

wobei hier und auch zukünftig über doppelt auftretende Indizes in Produkten stets summiert wird.

Es wird im nachfolgenden dieses Dualitätsprinzip auf Variationsprobleme in Parameterdarstellung unter Gebietsbeschränkungen angewandt und als Grundlage zur numerischen Abschätzung von $\inf_{\mathfrak{B}} J(x, u)$ benutzt.

2. Variationsprobleme in Parameterdarstellung

Wir spezialisieren die Aussage unseres Dualitätsprinzips (2) auf den Fall von Variationsproblemen unter Gebietsbeschränkungen (Zustandsbeschränkungen), aber ohne Steuerungsbeschränkungen. In Parameterdarstellung lautet somit unser Optimierungsproblem

$$J(x) = \int_0^T r(x, \dot{x})\, dt \to \text{Min} \quad \text{auf} \quad W_p^{1,n}(0, T) \tag{1'}$$

unter Gebietsbeschränkungen $x(t)\in\bar{G}_0 \ \forall t\in[0, T]$
und Randbedingungen $x(0) = x_0$, $x(T) = x_T$.

Im Sinne der Bezeichnungen von Abschnitt 1 ist jetzt r unabhängig von t, $g^i(t,\xi,v) \equiv v^i$, $l = n$, $V(t, \xi) \equiv E^n$ und $\bar{G} = \bar{G}_0 \times [0, T]$ mit $\bar{G}_0 \subset E^n$.

In üblicher Weise verlangen wir Invarianz von J gegenüber Parametertransformation bez. t; nach WEIERSTRASS bedeutet das: $r(x, v)$ soll bezüglich v positiv homogen ersten Grades sein (vgl. z. B. [2], Kap. 5). Darüber hinaus fordern wir $r(x, v) > 0 \ \forall v \neq 0$.

In diesem Spezialfall ist $H(\xi, v, y) = -r(\xi, v) + y_i v^i$, und wegen der positiven Homogenität ersten Grades von r bez. v ergibt sich jetzt

$$\sup_{v\in E^n} H(\xi, v, y) = \begin{cases} 0, & \text{falls} \quad -r(\xi, v) + y_i v^i \leqq 0 \quad \forall v\in E^n \\ \infty & \text{andernfalls.} \end{cases}$$

Unter Beschränkung auf t-unabhängige S, also auf $S \in D^1(\bar{G}_0)$, deren Gradient S_ξ der zusätzlichen Restriktion

$$S_\xi(\xi) \in \mathfrak{F}(\xi) := \{z \in E^n \mid z_i v^i \leqq r(\xi, v) \quad \forall v \in E^n\} \tag{3}$$

für alle $\xi \in \bar{G}_0$ genügt, wird somit $\delta_S(t, \xi) \equiv 0$, so daß die Dualitätsbeziehung (2) bez. (1') jetzt die einfache Gestalt annimmt:

$$J(x) \geqq S(x_T) - S(x_0) \quad \forall S \in \mathfrak{A}(G_0) \tag{4}$$

mit

$$\mathfrak{A}(G_0) = \{S \in D^1(\bar{G}_0) \mid S_\xi(\xi) \in \mathfrak{F}(\xi) \, \forall \xi \in \bar{G}_0\} .$$

Wie in [6] nachgewiesen wurde, gilt sogar

$$J(x) \geqq S(x_T) - S(x_0) \quad \forall S \in \bar{\mathfrak{A}}(G_0) \tag{4'}$$

mit $\bar{\mathfrak{A}}(G_0) = \{S \in W^1_\infty(G_0) \mid S_\xi(\xi) \in \mathfrak{F}(\xi)$ fast überall auf $G_0\}$.

Die Menge $\mathfrak{F}(\xi) \subset E^n$ ist übrigens im Sinne von Carathéodory, Minkowski und Hadamard der *Figuratrixkörper* des Variationsproblems (1'). Aus (3) entnehmen wir unmittelbar, daß $\mathfrak{F}(\xi)$ für jedes $\xi \in \bar{G}_0$ eine konvexe Menge des E^n ist, die wegen $r(\xi, v) > 0 \, \forall v \neq 0$ den Nullpunkt als inneren Punkt enthält.

Nach [6] gilt weiterhin in dem hier zugrunde gelegten Spezialfall

$$\inf_{\mathfrak{P}} J(x) = \operatorname*{Max}_{S \in \bar{\mathfrak{A}}(G_0)} [S(x_T) - S(x_0)] . \tag{5}$$

Daraus resultiert die prinzipielle Möglichkeit, mittels geeigneter Prozesse $\langle x, u \rangle$ und $S \in \mathfrak{A}(G_0)$ den Wert $I(x_0, x_T) = \inf_{\mathfrak{P}} J(x)$ durch $J(x)$ und $S(x_T) - S(x_0)$ beliebig genau von beiden Seiten einschließen zu können. Konstruktive Ausführungen dazu werden im nächsten Abschnitt aufbereitet.

3. Ein graphentheoretisches Näherungsprinzip

Wir betten in $\bar{G}_0$ einen Graphen Γ ein, bestehend aus endlich vielen Knotenpunkten $x_j \in \bar{G}_0$ $(j = 0, \ldots, N)$ und gewissen orientierten geradlinigen Verbindungskanten $\overrightarrow{x_i x_j}$, die vollständig zu $\bar{G}_0$ gehören sollen (*zulässige Verbindungskanten*). Der so gebildete gerichtete Graph Γ enthalte insbesondere auch x_0 und $x_T = x_N$ als Knotenpunkte, und es gebe zu jedem Knotenpunkt x_j mindestens einen (Kanten-)Weg $\mathfrak{W}(x_0, x_j)$ auf Γ, der den Anfangspunkt x_0 mit x_j verbindet. Den zulässigen Verbindungskanten $\overrightarrow{x_i x_j}$ ordnen wir die Bewertung

$$d_{ij} = \int_0^1 r(x_i + [x_j - x_i]\, t, x_j - x_i)\, dt > 0 \tag{6}$$

als „*Kantenlänge*" zu.

Zur näherungsweisen Berechnung von $I(x_0, x_T)$ bestimmen wir nun mittels bekannter elementarer Algorithmen (vgl. z. B. [1], [8] und [10]) auf dem kantenbewerteten Graphen Γ den kürzesten Weg von x_0 nach x_T; seine Länge bezeichnen wir mit $I_\Gamma(x_0, x_T)$. Die erwähnten Algorithmen liefern zugleich $I_\Gamma(x_k) := I_\Gamma(x_0, x_k)$ für alle Knotenpunkte x_k von Γ. Gemäß dieser Definition gilt offensichtlich

$$I_\Gamma(x_k) \leqq I_\Gamma(x_j) + d_{jk}\,, \tag{7}$$

falls von x_j nach x_k eine zulässige Verbindungskante existiert. Da die $\mathfrak{W}(x_0, x_T)$ außerdem spezielle (stückweise geradlinige) Trajektorien zu (1′) darstellen, deren Integralwert gemäß (6) auf jedem ihrer Streckenabschnitte $\overrightarrow{x_i x_j}$ gerade mit d_{ij} übereinstimmt, ist

$$I_\Gamma := I_\Gamma(x_0, x_T) \geqq I := I(x_0, x_T)\,.$$

Wir erhalten somit zunächst obere Schranken von I.

Zum Aufbau unterer Schranken benutzen wir nachfolgende prinzipielle Vorgehensweise. Wir verwenden die Werte $I_\Gamma(x_k)$ aus der vorangehenden Rechnung als Knotenbewertungen. Damit definieren wir vorerst nur in den Knotenpunkten x_k eine Funktion S mittels $S(x_k) := \alpha I_\Gamma(x_k)$ $(k=0, \ldots, N)$; α ist dabei eine noch näher zu bestimmende nichtnegative Konstante. Somit ist $S(x_0)=0$. Die Zahl α werde im weiteren so gewählt, daß sich S stetig auf $\bar{G}_0$ als Funktion von $\mathfrak{A}(G_0)$ bzw. $\tilde{\mathfrak{A}}(G_0)$ fortsetzen läßt; z. B. $\alpha=0$ gewährleistet das immer, weil $\mathfrak{F}(\xi)$ den Nullpunkt für alle $\xi \in \bar{G}_0$ enthält. Dann gilt offenbar gemäß (4) bzw. (4′)

$$I \geqq S(x_T) - S(x_0) = \alpha I_\Gamma(x_T) = \alpha I_\Gamma\,.$$

Zusammenfassend ergibt diese Überlegung mit dem obigen Resultat die *beidseitige Abschätzung von I*:

$$I_\Gamma \geqq I \geqq \alpha I_\Gamma\,. \tag{8}$$

Aus dieser Abschätzung heraus sehen wir unmittelbar, daß $\alpha \leqq 1$ sein muß. Andererseits wird man bemüht sein, um mit (8) eine möglichst enge Einschließung von I zu erhalten, die Konstante α möglichst groß aus $[0, 1]$ zu wählen. Wir werden im nächsten Abschnitt und in einer Nachfolgearbeit gleichen Titels (Teil II) praktikable Berechnungsvorschriften für solche α aufstellen.

4. Methode 1

$\bar{G}_0$ sei zunächst ein polyedrischer Bereich des E^n. Γ werde aus sämtlichen Eckpunkten (Knotenpunkten) und Kanten einer Triangulierung von $\bar{G}_0$ in ϱ reguläre n-Simplexe $\Delta_i (i=1, \ldots, \varrho)$ gebildet. Wiederum seien x_0 und x_T in der Menge aller Knotenpunkte $x_j (j=0, \ldots, N)$ enthalten. Wir berechnen dann gemäß vorangehendem Abschnitt sämtliche Knotenbewertungen $I_\Gamma(x_k)$. Damit definieren wir eine Funktion S vorerst nur auf den Knotenpunkten x_k durch den Ansatz $S(x_k) := \alpha I_\Gamma(x_k)$ $(k=0, \ldots, N)$ mit $\alpha \in [0, 1]$. Anschließend bestimmen wir auf jedem Δ_i jene lineare Erweiterung von S, die durch die Bedingung

$$S(x_k) = \alpha I_\Gamma(x_k) \quad \forall x_k \in \Delta_i \tag{9}$$

eindeutig definiert ist und in Δ_i die Gestalt $S(x)=\alpha L_i(x)$ besitzt; L_i ist dabei eine in Δ_i definierte lineare Funktion. Indem wir uns diese Prozedur für jedes Δ_i ausgeführt denken, erhalten wir damit auf ganz $\bar{G}_0$ eine stückweise (auf den einzelnen Δ_i) lineare Funktion S, die offensichtlich stetig ist und zu $D^1(\bar{G}_0)$ gehört.

Um zu erreichen, daß auch $S\in\mathfrak{A}(G_0)$ gilt, genügt es, das α so zu wählen, daß für sämtliche $\Delta_i\,(i=1,\ldots,\varrho)$

$$\alpha L_{ix}\in\bigcap_{\xi\in\Delta_i}\mathfrak{F}(\xi)=\mathfrak{F}_i \tag{10}$$

erfüllt ist (man beachte, daß der Gradient L_{ix} von L_i auf Δ_i ein konstanter Vektor ist). Die Bedingung (10) ist nach (3) äquivalent zur Forderung

$$\alpha L_{i\xi^j}(\xi)v^j\leqq r_i(v):=\operatorname*{Min}_{\xi\in\Delta_i} r(\xi,v)\qquad\forall v\in E^n\,. \tag{10'}$$

Da wir bemüht waren, für (8) ein möglichst großes α zu wählen, empfiehlt sich nach (10') die folgende Festlegung:

Setze $\alpha=\operatorname*{Min}_i\alpha_i$ mit $i=1,\ldots,\varrho$ und

$$\alpha_i^{-1}=\operatorname*{Max}_{v\neq0}\,[L_{i\xi^j}v^j/r_i(v)]\,. \tag{11}$$

Mit diesem α gilt dann offenbar Abschätzung (8).

Diese über (11) angegebene Berechnung zulässiger α setzt allerdings real die Kenntnis der $I_\Gamma(x_k)$ $(k=0,\ldots,N)$ voraus. Man kann jedoch auf nachstehende Weise auch a priori zulässige α im Sinne von (10) berechnen, ohne vorangehend die $I_\Gamma(x_k)$ numerisch wirklich bestimmt zu haben. Dazu betrachten wir für jedes feste Δ_i seine $n+1$ Eckpunkte $x_{k_{0(i)}},\ldots,x_{k_{n(i)}}$ – wir lassen im weiteren zur Vereinfachung der Schreibweise die Klammern (i) wieder weg. Da Δ_i konvex ist, hat jeder Punkt $\xi\in\Delta_i$ die eindeutige Darstellung

$$\xi=\sum_{\nu=0}^{n}\lambda_\nu x_{k_\nu}\quad\text{mit}\quad\lambda_\nu\geqq0,\quad\sum_{\nu=0}^{n}\lambda_\nu=1\,.$$

Wegen der Linearität von L_i auf Δ_i gilt nach (9)

$$L_i(\xi)=\sum_{\nu=0}^{n}\lambda_\nu I_\Gamma(x_{k_\nu})\,.$$

Zugleich bildet bei festgehaltenem $x_{k_\mu}\in\Delta_i$ das Vektorsystem $(x_{k_\nu}-x_{k_\mu})$ $\forall x_{k_\nu}\in\Delta_i$, $\nu\neq\mu$, eine Basis des E^n, so daß zu jedem Vektor $v\in E^n$ eindeutig eine Darstellung $v=a_\nu\,(x_{k_\nu}-x_{k_\mu})$ existiert. Unter Einführung der Einheitsvektoren

$$\mathfrak{e}_\nu=\begin{cases}(x_{k_\nu}-x_{k_\mu})/|x_{k_\nu}-x_{k_\mu}| & \text{für}\quad a_\nu\geqq0\\(x_{k_\mu}-x_{k_\nu})/|x_{k_\nu}-x_{k_\mu}| & \text{für}\quad a_\nu<0\,,\end{cases} \tag{12}$$

die in bezug auf v genau n ausgezeichneten Kantenrichtungen von Δ_i entsprechen, hat mit $\beta^\nu=|a_\nu|\,|x_{k_\nu}-x_{k_\mu}|$ der Vektor v die Darstellung $v=\beta^\nu\mathfrak{e}_\nu$ mit $\beta^\nu\geqq0$. Lassen wir an die Stelle von v speziell einen Einheitsvektor $\mathfrak{e}$ treten, so können wir nach

Vorangehendem von der Existenz einer solchen zugeordneten Basis von Einheitsvektoren (12) ausgehen, daß $\mathfrak{e}$ eine Darstellung

$$\mathfrak{e}=\beta^\nu \mathfrak{e}_\nu \quad \text{mit} \quad \beta^\nu \geqq 0 \tag{13}$$

besitzt; im übrigen läßt sich β^ν mittels der *reziproken Basis* $\{\mathfrak{e}^\nu\}$ zu $\{\mathfrak{e}_\nu\}$ auch als das Skalarprodukt $\beta^\nu=(\mathfrak{e}, \mathfrak{e}^\nu)$ schreiben (vgl. z. B. [9], S. 274).

Mit diesen Vorbetrachtungen können wir wegen der positiven Homogenität ersten Grades von r_i die Bedingung (10′) zunächst durch die äquivalente Forderung

$$\alpha L_{i\xi^j} v^j \leqq r_i(v) \quad \forall v \in E^n \quad \text{mit} \quad |v|=1$$

ersetzen. Mit $v=\mathfrak{e}$ stellt die linke Seite dieser Ungleichung gerade das Produkt von α mit der Richtungsableitung $\frac{\partial L_i}{\partial \mathfrak{e}}$ dar. Unter Verwendung von (13) gilt dabei $\frac{\partial L_i}{\partial \mathfrak{e}}=\frac{\partial L_i}{\partial \mathfrak{e}_\nu}\cdot(\mathfrak{e}, \mathfrak{e}^\nu)$. Die Richtungsableitungen $\frac{\partial L_i}{\partial \mathfrak{e}_\nu}$ haben nach Definition von L_i und $\mathfrak{e}_\nu$ den Wert

$$\frac{\partial L_i}{\partial \mathfrak{e}_\nu}=\begin{cases}[I_\Gamma(x_{k_\nu})-I_\Gamma(x_{k_\mu})]/|x_{k_\nu}-x_{k_\mu}| & \text{für} \quad a_\nu \geqq 0\\ [I_\Gamma(x_{k_\mu})-I_\Gamma(x_{k_\nu})]/|x_{k_\nu}-x_{k_\mu}| & \text{für} \quad a_\nu < 0.\end{cases} \tag{14}$$

Deshalb ist (10′) dann und nur dann erfüllt, wenn für sämtliche Einheitsvektoren $\mathfrak{e}\in E^n$ und zugeordnete Kantenbasen $\{\mathfrak{e}_\nu\}$ zu Δ_i im Sinne von (12) die Bedingung

$$\alpha \frac{\partial L_i}{\partial \mathfrak{e}}=\alpha\cdot(\mathfrak{e}, \mathfrak{e}^\nu)\frac{\partial L_i}{\partial \mathfrak{e}_\nu}\leqq r_i(\mathfrak{e}) \tag{10''}$$

gilt. Wegen (7) und $(\mathfrak{e}, \mathfrak{e}^\nu)\geqq 0$ ist (10″) garantiert erfüllt, wenn sogar folgende Eigenschaft besteht:

$$\alpha(\mathfrak{e}, \mathfrak{e}^\nu)\,\delta_{k_\mu k_\nu}/|x_{k_\nu}-x_{k_\mu}| \leqq r_i(\mathfrak{e}) \tag{15}$$

$$\text{mit} \quad \delta_{k_\mu k_\nu}=\begin{cases} d_{k_\mu k_\nu} & \text{für} \quad a_\nu \geqq 0\\ d_{k_\nu k_\mu} & \text{für} \quad a_\nu < 0\,.\end{cases}$$

In dieser Bedingung spielt die Kenntnis der $I_\Gamma(x_k)$ keine Rolle mehr.

Zusammenfassend ergibt diese Studie folgendes Resultat.

Satz 1: *Zum Variationsproblem* (1′) *sei der Zustandsbereich* $\bar{G}_0$ *triangulierbar in* ϱ *reguläre n-Simplexe* Δ_i. *Dann gilt für den Minimalwert I* (*bzw. das Infimum*) *Abschätzung* (8), *wenn die Konstante* α *über die Vorschrift* (11) *gewählt wird oder nachstehenden Bedingungen genügt*:

Zu jedem $\Delta_i (i=1, \ldots, \varrho)$ *und Einheitsvektor* $\mathfrak{e}\in E^n$ *mit* $\mathfrak{e}=a_\nu(x_{k_\nu}-x_{k_\mu})$ *bez. sämtlicher Eckpunkte* x_{k_ν} *von* Δ_i *und willkürlich gewähltem* (*festen*) *Eckpunkt* x_{k_μ} *von* Δ_i *gelte Bedingung* (15).

Dieser analytischen Berechnungsvorschrift von α können wir sehr leicht auch ein geometrisches Analogon gegenüberstellen, wenn wir die geometrische Bedeutung von (10) und (14) voll erfassen. Beachten wir nämlich, daß $\frac{\partial L_i}{\partial \mathfrak{e}_\nu}$ den Abstand des Projektionspunktes von L_{ix} auf den $\mathfrak{e}_\nu$-Strahl vom Koordinatenursprung O angibt, so ist nach (14) und (7) L_{ix} Element eines Polyeders $\mathfrak{K}_1^{(i)} \subset E^n$. Dieses ist Durchschnitt aller jener Halbräume $H_{\mu\nu}^+$, die den Koordinatenursprung O enthalten und von Hyperebenen $H_{\mu\nu}$ begrenzt werden, welche orthogonal zum Vektor $x_{k_\nu} - x_{k_\mu}$ verlaufen und den Abstand

$$z_{\mu\nu} := d_{k_\mu k_\nu} / |x_{k_\nu} - x_{k_\mu}| \tag{16}$$

vom Ursprung O haben. μ und ν ($\mu \neq \nu$) durchlaufen dabei sämtliche Indizes mit $x_{k_\mu}, x_{k_\nu} \in \Delta_i$.

Wählen wir hingegen statt jenes Abstandes (16) den Wert $\alpha z_{\mu\nu}$, so liefert vorangehende Konstruktion ein zu $\mathfrak{K}_1^{(i)}$ ähnliches Polyeder $\mathfrak{K}_\alpha^{(i)}$. Treffen wir die Wahl des $\alpha > 0$ schließlich so klein, daß für alle $i = 1, \ldots, \varrho$ die Inklusionen $\mathfrak{K}_\alpha^{(i)} \subset \mathfrak{F}_i$ gelten, so ist wegen $L_{ix} \in \mathfrak{K}_1^{(i)}$ zugleich $\alpha L_{ix} \in \mathfrak{K}_\alpha^{(i)} \subset \mathfrak{F}_i$, also Bedingung (10) resp. (10′) erfüllt.

In der Zusammenfassung lautet dieses geometrische Resultat:

Satz 2: *Unter folgender Konstruktionsvorschrift für α gilt die Abschätzung* (8):

1. *Trage vom Koordinatenursprung O des E^n aus sämtliche Strahlen $s_{\mu\nu}$ der Richtung $x_{k_\nu} - x_{k_\mu}$ an, wobei die $x_{k_\nu} = x_{k_\nu(i)}$ die Eckpunkte von Δ_i bei fixiertem i sind.*

2. *Konstruiere die Halbräume $H_{\mu\nu}^{(i)}$, welche O enthalten und von Hyperebenen begrenzt werden, welche orthogonal zu $s_{\mu\nu}$ verlaufen und den Abstand $\alpha z_{\mu\nu}$ gemäß* (16) *von O haben.*

3. *Bilde für jedes $i = 1, \ldots, \varrho$ den Durchschnitt $\mathfrak{K}_\alpha^{(i)} = \bigcap_{\mu,\nu} H_{\mu\nu}^{(i)}$ aller dieser Halbräume und wähle α so klein, daß $\mathfrak{K}_\alpha^{(i)} \subset \mathfrak{F}_i$ gilt $\forall i = 1, \ldots, \varrho$.*

Beispiel: Der kürzeste Weg zwischen zwei Punkten eines ebenen Labyrinths (in euklidischer Maßbestimmung).

Wir setzen vereinfachend voraus, daß das Labyrinth $\bar{G}_0 \subset E^2$ so beschaffen sei, daß es eine Triangulierung in kongruente gleichseitige Dreiecke Δ_i gestatte ($i = 1, \ldots, \varrho$). Sämtliche Δ_i sollen eine Seite parallel zur ξ^1-Achse eines zugrunde gelegten rechtwinkligen kartesischen (ξ^1, ξ^2)-Koordinatensystems besitzen. Wie vorangehend nehmen wir weiterhin an, daß Anfangspunkt x_0 und Endpunkt x_T Eckpunkte einiger dieser Dreiecke sind.

Im vorliegenden Beispiel ist $r(\xi, v) \equiv |v|$, $n = 2$ und – wie man leicht ersieht – $\mathfrak{F}(\xi) = \mathfrak{F}_i$ für alle i gleich dem Einheitskreis-Bereich zum Mittelpunkt O. Gemäß (6) und (16) sind hier sämtliche $z_{\mu\nu} = 1$. Im Sinne des geometrischen Konstruktionsprinzips von Satz 2 läuft die Konstruktion von α gemäß Fig. 1 jetzt darauf hinaus, ein reguläres Sechseck mit dem Mittelpunkt O und dem Seitenabstand α von O zu konstruieren, welches vom Einheitskreis umschlossen wird. Wir erreichen dies mit dem größtmöglichen Wert $\alpha = (1/2)\sqrt{3}$.

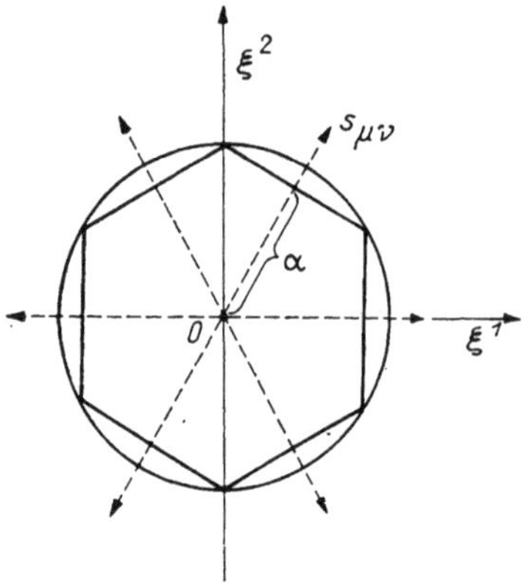

Fig. 1

Nach (8) gilt somit für unser Beispiel die A-priori-Abschätzung $I_\Gamma \geqq I \geqq (1/2)\sqrt{3}\, I_\Gamma$. Das bedeutet zugleich, daß der Näherungswert $\tilde{I} = (1/2)\,(I_\Gamma + (1/2)\sqrt{3}\, I_\Gamma)$ einen relativen Fehler von maximal 7% gegenüber I aufweist, unabhängig davon, wie die spezielle geometrische Struktur von G_0 beschaffen ist.

Wir bemerken, daß die errechneten Schranken dieser Einschließung von I in dem Sinne scharf sind, daß man stets Gebiete G_0 angeben kann, wo $I_\Gamma = I$ bzw. $(I = 1/2)\sqrt{3}\, I_\Gamma$ ist. Vgl. dazu die beiden in Fig. 2a bzw. 2b skizzierten Spezialfälle. In Fig. 2a ist $I_\Gamma = I$, und in Fig. 2b ist $I = (1/2)\sqrt{3}\, I_\Gamma$.

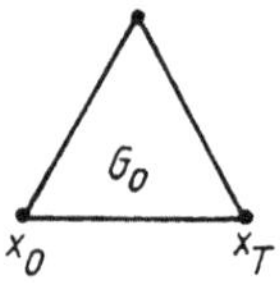

Fig. 2a

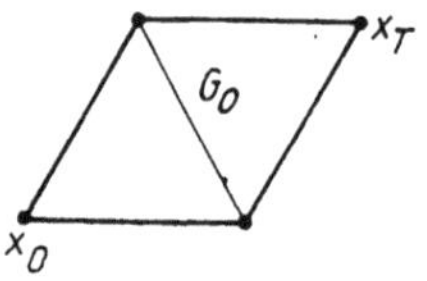

Fig. 2b

5. Erweiterungen

Die vorangehend begründeten Abschätzungsmethoden zu I bez. des Variationsproblems (1′) bleiben auch dann noch gültig, wenn gewisse zusätzliche Richtungsbeschränkungen (Steuerungsbeschränkungen) auftreten. Ohne wesentlich neue Gedankengänge kann man leicht zeigen, daß diese Behauptung insbesondere für Steuerungsprobleme des Typs

$$J(x) = \int_0^T r(x, \dot{x})\, dt \to \text{Min} \quad \text{auf} \quad W_p^{1,n}(0, T) \tag{1''}$$

unter Gebietsbeschränkungen $x(t) \in \bar{G}_0 \;\; \forall t \in [0, T]$,
Steuerungsbeschränkungen $\dot{x}(t) \in V$ und
Randbedingungen $x(0) = x_0$, $x(T) = x_T$
richtig ist, wenn dieselben Voraussetzungen wie zu (1′) erfüllt sind und V ein konstanter abgeschlossener Kegel des E^n mit der Spitze in O ist.

Rein formal geschieht die Übertragung der Abschätzungstechnik von Abschnitt 3–4 für das $\inf J(x)$ bez. (1″) dadurch, daß wir (1″) in ein Problem ohne Steuerungsbeschränkungen umwandeln. Wir führen dazu den neuen Integranden

$$\hat{r}(\xi, v) = \begin{cases} r(\xi, v) & \forall v \in V \\ \infty & \forall v \notin V \end{cases}$$

ein und behandeln statt (1″) das äquivalente Variationsproblem (1′) mit dem Integranden $\hat{r}$. Sofern sämtliche $I_\Gamma(x_j)$ $(j = 0, \ldots, N)$ endlich sind, behalten alle Entwicklungen der Abschnitte 3–4 auch jetzt ihre volle Gültigkeit, selbst wenn einige der $d_{ij} = \infty$ sind. $\mathfrak{F}(\xi)$ wird freilich jetzt i. allg. nicht mehr beschränkt sein.

Die „Methode 1" von Abschnitt 4 zum Aufbau einer Abschätzung vom Typ (8) war zunächst an die wesentliche Voraussetzung gebunden, daß $\bar{G}_0$ triangulierbar und somit polyedrisch ist. In welcher Weise können wir mit unseren Methoden auch Abschätzungen von I gewinnen, wenn G_0 ein allgemeineres Normalgebiet des E^n ist? Die Antwort lautet:

Konstruiere einen polyedrischen Bereich $\bar{G}' \supset \bar{G}_0$ und trianguliere ihn. x_0 und x_T seien wiederum Eckpunkte von Simplexen dieser Triangulierung. Auf dem durch die Triangulierung von G' erzeugten Graphen Γ' berechne man $I_{\Gamma'} = I_{\Gamma'}(x_0, x_T)$ und ein α nach Satz 1 bzw. 2. Das damit erzeugte $S \in \mathfrak{A}(G')$ ist also erst recht Element von $\mathfrak{A}(G_0)$. Daraus folgt $I \geqq \alpha I_{\Gamma'}$ nach (4).

Zum anderen betrachten wir die Gesamtheit aller Simplexe Δ_i von $\bar{G}'$, die vollständig zu $\bar{G}_0$ gehören. Ihre Vereinigungsmenge $\bar{G}'' \subset \bar{G}_0$ ist nach Konstruktion bereits trianguliert. Auf dem durch die Triangulierung von $\bar{G}''$ erzeugten Graphen Γ'' berechne man $I_{\Gamma''} = I_{\Gamma''}(x_0, x_T)$. Da sämtliche Wege auf Γ'' offensichtlich spezielle Trajektorien von $\bar{G}_0$ sind, gilt $I_{\Gamma''} \geqq I$. Somit tritt an die Stelle von (8) bei nicht polyedrischen Bereichen $\bar{G}_0$ die Einschließung

$$I_{\Gamma''} \geqq I \geqq \alpha I_{\Gamma'}.$$

Die beschriebene Methode 1 ist konstruktiv relativ einfach, sie hat jedoch den Nachteil (wie schon das Beispiel zeigt), in sich keine Verbesserung der Einschließung von I zu gestatten, die Schranken bleiben noch ziemlich grob. Die Ursache dafür liegt darin, daß der gewählte Graph bei Methode 1 an die Triangulierung von $\bar{G}_0$ gebunden wurde. D. h., als Kanten wurden nur Verbindungsstrecken von Eckpunkten gleicher Simplexe zugelassen. Die Güte der Einschließung (8) kann gegenüber Methode 1 erheblich verbessert werden, wenn man die Anzahl der zugelassenen Verbindungskanten zwischen Eckpunkten erhöht. Die Ausführung dieser Idee bleibt einer Nachfolgearbeit vorbehalten.

Literatur

[1] Berge, C., und A. Ghouila-Houri: Programme, Spiele, Transportnetze, Leipzig 1969.
[2] Bolza, O.: Vorlesungen über Variationsrechnung. Leipzig 1949.
[3] Klötzler, R.: Einige neue Aspekte zur Bellmanschen Differentialgleichung. Материалы всесоюзного симпозиума по оптимальному управлению и дифференциальным играм, Тбилиси 1976, 146–154.

[4] Klötzler, R.: Weiterentwicklungen der Hamilton-Jacobischen Theorie. Forschungsberichte d. Akad. d. Wiss. d. DDR (im Druck).
[5] Klötzler, R.: On a general conception of duality in optimal control. Proceedings of the conference EQUADIFF 4, Prague 1977 (im Druck).
[6] Klötzler, R.: A generalization of the duality in optimal control and some numerical conclusions. Proceedings of the 8th IFIP Conference, Würzburg 1977, 313–320.
[7] Klötzler, R.: Mehrdimensionale Variationsrechnung. Berlin, 1969.
[8] Krekó, B.: Optimierung nichtlinearer Modelle. Berlin 1974.
[9] Lagally, M.: Vorlesungen über Vektorrechnung. Leipzig 1949.
[10] White, D. J.: Dynamic Programming. Edinburgh, London 1969.

Summary

The present paper deals with an application of a new general duality principle to parametric variational problems for simple integrals on state restrictions. It is developed a both-sided estimate of the optimal value and its numerical realization by means of graph theory methods.

Резюме

В работе применяется установленный автором ранее новый общий принцип двойственности к вариационным проблемам одномерных интегралов в параметрическом представлении фазовыми ограничениями. Выводится двусторонная оценка для оптимального значения и его численная реализация методом теории графов.

Eingereicht bei der Redaktion: 30. 1. 1978

Math. Operationsforsch. Statist., Ser. Optimization, Vol. 10 (1979) No. 1, 111–119

Über Werteannahme bei einer Klasse von Optimalproblemen[1]

A. Göpfert[2]

Zusammenfassung: Für gewisse Optimalprobleme vom Approximationstyp ohne Annahme des Minimalwertes muß das Dualproblem instabil sein. Als analytische Ursache der Instabilität erweisen sich gerade gestörte, inkorrekt gestellte Randanfangswertprobleme. Die Minimalwertannahme wird durch eine Bedingung charakterisiert, die aus der Stabilitätstheorie der Optimierung folgt.

1. Einführung

Für eine gegebene konvexe Optimierungsaufgabe

$$\{F(x) \mid x \in M\} = \text{Min!} \tag{1}$$

sei der Optimalwert endlich. Er werde aber nicht angenommen. Im Rahmen der durch eine Störung von (1) zusammen mit dem Konzept der konjugierten Funktionen erzeugten Dualitätstheorie ([7], [8]) drückt sich der Sachverhalt der Nicht-Werteannahme notwendig in einer Instabilität des Dualproblems gegenüber (dualen) Störungen aus: die Optimalwertfunktion $S(z^*)$ des gestörten Dualproblems (Störung: z^*, $z^* \in U^* \subset Z^*$, U^* kleine konvexe Nullumgebung im Störraum $\equiv$ $\equiv$ Parameterraum Z^*) kann an der Stelle $z^* = 0$ nicht gleichmäßig $\forall z^* \in U^*$ nach oben beschränkte Richtungsableitungen haben. Kleine Änderungen (im Sinne der gewählten Störung) der Daten des Dualproblems können also nicht nur kleine Anstiege der Änderungen des Optimalwertes nach sich ziehen. Im einfachen Beispiel ([4], S. 122)

$$\{2x^{-1} \mid x \in [1, \infty)\} = \text{Min!} \tag{2}$$

eines Optimierungsproblems ohne Werteannahme ist der Infimalwert $J = 0$, er wird nicht angenommen, der Maximalwert des Fenchelschen Dualproblems ist $S = 0$.

---

[1] Vortrag, gehalten auf der Tagung „Mathematische Optimierung – Theorie und Anwendungen“ in Eisenach vom 5.–10. 12. 1977.

[2] Technische Hochschule „Carl Schorlemmer“ Leuna-Merseburg, Sektion Mathematik und Rechentechnik, DDR-42 Merseburg, Geusaer Str.

Wird das Dualproblem „horizontal“ ([7], S. 24) gestört:

$$S(z^*)=\operatorname{Sup}\{g^*(x^*)-f^*(x^*-z^*)\}, \tag{3}$$

wobei f^* und g^* die konjugierten Funktionen zu $f(x)=x^{-1}$, $\operatorname{dom} f=[1,\infty)$, $g(x)=-x^{-1}$, $\operatorname{dom} g=[1,\infty)$ sind und $z^*\in U^*$ mit $U^*=\{z^*\in R^1 \mid |z^*|<1\}$ ist, so ergibt sich

$$S(z^*)=4\sqrt{\frac{z^*}{2}} \quad \text{für} \quad 0\leqq z^*<1 \tag{4}$$

und $S(z^*)=-\infty$ für $z^*<0$. Offenbar sind die Richtungsableitungen unbeschränkt nach oben für Richtungen $z^*>0$.

Wir betrachten im folgenden Steuerprobleme des Typs

$$\{\|a-Ax\| \mid x\in \boldsymbol{X}\}=\text{Min!} \tag{5}$$

mit Kern $N(A^*)=\{0\}$ und charakterisieren einerseits die Erreichbarkeitsmenge $\{Ax \mid x\in \boldsymbol{X}\}$ durch die Stabilität des Dualproblems und studieren andererseits die Ursache der Instabilität für den Fall, daß A GREENscher Integraloperator ist. Dabei ergibt sich interessanterweise, daß gerade das klassische instabile Verhalten nicht korrekt im HADAMARDschen Sinne gestellter Randanfangswertprobleme die Instabilität im Sinne der Dualitätstheorie verursacht.

2. Primal- und Dualproblem

Es seien $\boldsymbol{X}$ und $\boldsymbol{Y}$ BANACHräume, $A: \boldsymbol{X}\to \boldsymbol{Y}$ ein linearer stetiger Operator von $\boldsymbol{X}$ in $\boldsymbol{Y}$ mit Nullraum von A^*: $N(A^*)=\{0\}$. $\boldsymbol{X}^*$ und $\boldsymbol{Y}^*$ seien die Dualräume, so daß $(\boldsymbol{X} \mid \boldsymbol{X}^*)$, $(\boldsymbol{Y} \mid \boldsymbol{Y}^*)$ duale Paare sind. $a\in \boldsymbol{Y}$ sei ein vorgegebenes Element. Die Aufgabe lautet, a durch Ax mit $x\in \boldsymbol{X}$ im Sinne der Norm in $\boldsymbol{Y}$ möglichst genau zu approximieren:

$$\{\|a-Ax\|_{\boldsymbol{Y}} \mid x\in \boldsymbol{X}\}=\text{Min!}. \tag{5'}$$

Beispiel: ∂E sei der durch reell-analytische Funktionen beschreibbare Rand eines der linearen Elastizitätstheorie gehorchenden elastischen Körpers E. Über ∂E betrachten wir den Banachraum $\boldsymbol{X}$ der stetigen Funktionen (oder den Hilbertraum $\boldsymbol{X}$ der L_2-Funktionen). Im Innern von E sei ein glattes Flächenstück Γ, das E nicht zerlegt, gegeben. Über Γ betrachten wir den BANACHraum $\boldsymbol{Y}$ der stetigen Funktionen (oder den HILBERTraum $\boldsymbol{Y}$ der L_2-Funktionen). a sei ein über Γ gegebener Verrückungsvektor. Ax sei die durch das elliptische 1. Randwertproblem der linearen Elastizitätstheorie

$$\begin{aligned} \Delta^* u=\Delta u+\frac{\lambda+\mu}{\mu}\operatorname{grad}\operatorname{div} u=0 \quad &\text{in} \quad \operatorname{int} E, \\ u=x \quad \text{auf} \quad \partial E& \end{aligned} \tag{6}$$

eindeutig existierende Lösung, eingeschränkt auf Γ. λ und μ sind die LAMÉschen Konstanten, $\lambda+2\mu>\frac{4}{3}\mu>0$. Hat Γ eine zweimal stetig differenzierbare Darstellung, so ist eine Lösung über Γ auch zweimal stetig differenzierbar. Ist daher der auf Γ vorgegebene Verrückungsvektor a etwa nicht differenzierbar auf einer Teilmenge von Γ, so kann keine Werteannahme eintreten, das Dualproblem muß instabil sein. $N(A^*)=\{0\}$ folgt in Abschnitt 4.

Wir betrachten nun für $\boldsymbol{Y}=C(\Gamma)$, $\boldsymbol{X}=C(\partial E)$, wobei Γ und ∂E unabhängig vom Problem (6) nur die Trägermengen der Räume $\boldsymbol{X}$ und $\boldsymbol{Y}$ sein mögen[3], das im ROCKAFELLARschen Sinne duale Problem zu (5'). (Für $\boldsymbol{Y}=L_2(\Gamma)$, $\boldsymbol{X}=L_2(\partial E)$ wurde es in [5] berechnet.) Diese Bildung des dualen Problems entspricht der horizontalen Störung in [7]. Äquivalent zu (5') ist

$$\{\xi \mid \xi\in R^1, \xi\geqq 0, \exists x\in \boldsymbol{X}: |Ax-a|\leqq\xi \quad \text{über} \quad \Gamma, A:\boldsymbol{X}\to\boldsymbol{Y}\}=\text{Min!}. \tag{7}$$

Es seien $v=(x, \xi)^{\mathsf{T}}$ die Elemente des Produktraumes $\boldsymbol{V}=\boldsymbol{X}\times R^1$, $v^*=(x^*, \xi)^{\mathsf{T}}$ die Elemente von $\boldsymbol{X}^*\times R^1$, ferner $\boldsymbol{A}$ die lineare stetige Abbildung

$$\boldsymbol{A}=\begin{pmatrix} A & I \\ -A & I\end{pmatrix}: \boldsymbol{V}\to\boldsymbol{W} \tag{8}$$

mit $\boldsymbol{A}v=(Ax+\xi, -Ax+\xi)^{\mathsf{T}}$, wobei $(g_1, g_2)^{\mathsf{T}}$ die Elemente in $\boldsymbol{W}=\boldsymbol{Y}\times\boldsymbol{Y}$ seien.[4] In $\boldsymbol{X}^*\times R^1$, $\boldsymbol{Y}^*\times\boldsymbol{Y}^*$ sei die schwache *-Topologie (oder eine stärkere, höchstens aber die MACKEY-Topologie) gewählt. Mit den Kegel-Halbordnungen

$$K_{\boldsymbol{W}}=\{(g_1, g_2)^{\mathsf{T}} \mid g_1, g_2\in\boldsymbol{Y}, g_i\geqq 0 \quad \text{über} \quad \Gamma, i=1, 2\}, \tag{9}$$

$$K_{\boldsymbol{V}}=\{v=(x, \xi)^{\mathsf{T}} \mid x\in\boldsymbol{X}, \xi\geqq 0\} \tag{10}$$

und

$$P^{\mathsf{T}}=(0, 1)^{\mathsf{T}}\in\boldsymbol{V}, \quad h=(a, -a)^{\mathsf{T}}\in\boldsymbol{W} \tag{11}$$

hat dann (7) die Gestalt eines gewöhnlichen linearen Optimalproblems:

$$\boldsymbol{A}v\underset{K_{\boldsymbol{W}}}{\geqq}h, \quad v\underset{K_{\boldsymbol{V}}}{\geqq}0, \quad \langle p, v\rangle=\text{Min!}. \tag{12}$$

Sein Dualproblem wird wie in der linearen Optimierung gebildet, wobei die adjungierten Halbordnungen und Abbildungen auftreten:

$$\boldsymbol{A}^*w^*\underset{K^*_{\boldsymbol{V}}}{\leqq}p, \; w^*\underset{K^*_{\boldsymbol{W}}}{\geqq}0, \quad \langle w^*, h\rangle=\text{Max!}. \tag{12D}$$

Wegen $K^*_{\boldsymbol{V}}=\{0\}\times R^1_+$, $A^*: \boldsymbol{Y}^*\to\boldsymbol{X}^*$, $I^*: \boldsymbol{Y}^*\to R^1$ und

$$\boldsymbol{A}^*=\begin{pmatrix} A^* & -A^* \\ I^* & I^*\end{pmatrix}$$

bekommt (12D) seine aus der Approximationstheorie bekannte Form, wobei noch $w^*=(w_1^*, w_2^*)^{\mathsf{T}}$ gesetzt werde:

$$\begin{aligned} &\langle w_1^*-w_2^*, a\rangle=\text{Max!}, \quad w_1^*\geqq 0, \quad w_2^*\geqq 0 \quad \text{in} \quad \boldsymbol{Y}^*, \\ &A^*(w_1^*-w_2^*)=0 \quad \text{in} \quad \boldsymbol{X}^*, \quad \langle w_1^*+w_2^*, 1\rangle\leqq 1 . \end{aligned} \tag{13}$$

---

[3] Als $\boldsymbol{X}$ könnte auch die Menge der stetigen finiten Funktionen genommen werden über einem kompakten Bereich des R^n.

[4] Der Operator I aus (8) bildet also R^1 in $\boldsymbol{Y}=C(\Gamma)$ ab, $I\xi$ ist diejenige Funktion, die konstant gleich ξ ist.

Da die w_i^* nichtnegativen Borel-Maßen isometrisch sind, ist die letzte Ungleichung gleichbedeutend mit $\|w_1^* - w_2^*\|_{Y^*} \leqq 1$.

Gerade ein Problem derselben Form wie (13) erhält man bei Dualisierung im quadratischen Fall, wobei der Dualisierungsformalismus dem Vorgehen bei einer Eisenbergschen Aufgabe entspricht. Indem in (13) $w_1^* - w_2^* = \mathfrak{w}^* \in \boldsymbol{Y}^*$ gesetzt wird, gilt

$$\langle \mathfrak{w}^*, a\rangle = \text{Max}!, \quad A^*\mathfrak{w}^* = 0, \quad \|\mathfrak{w}^*\| \leqq 1 . \tag{13'}$$

3. Stabilität des Primalproblems

Wir beweisen, daß keine Dualitätslücke vorhanden ist. Dazu zeigen wir zuerst, daß der Wertevorrat des Operators A dicht liegt in $\boldsymbol{Y}$. Denn gäbe es ein von Null verschiedenes Element $h^* \in \boldsymbol{Y}^*$ aus dem Annihilator von A, also

$$\langle h^*, Ax\rangle = 0 \quad \forall x \in \boldsymbol{X} ,$$

so folgt

$$\langle A^*h^*, x\rangle = 0 \quad \forall x \in \boldsymbol{X}$$

und hieraus $A^*h^* = 0$. Da $N(A^*) = \{0\}$ nach Voraussetzung, ist $h^* = 0$. Daher ist der Infimalwert J von (12): $J = 0$. Der Supremalwert S von (13′) ist aber wegen $N(A^*) = \{0\}$ gleich 0 und wird von $\mathfrak{w}^* = 0$ angenommen. $J = S$ mit Annahme von S bedeutet nach der Dualitätstheorie (schwacher Dualitätssatz), daß das Primalproblem stabil ist (nach [8]) oder „inf-dif-stable" (weil J endlich ist) nach [7]. Dieses Resultat bedeutet vom Standpunkt der Approximationsaufgabe aus: die Approximation von a ist beliebig genau möglich. Die Form der Störungen im Sinne der Dualitätstheorie erkennt man, wenn (12) in der Fenchelschen Form geschrieben wird:

$$\{k_1(v) - k_2(\boldsymbol{A}v) \mid v \in \boldsymbol{V}\} = \text{Inf}! , \tag{14}$$

wobei gilt

$$k_1(v) = \begin{cases} \langle p, v\rangle & \text{für} \quad v \underset{K_V}{\geqq} 0 \\ +\infty & \text{sonst} \end{cases} \tag{14.1}$$

$$k_2(w) = \begin{cases} 0 & \text{für} \quad w \underset{K_W}{\geqq} h \\ -\infty & \text{sonst.} \end{cases} \tag{14.2}$$

Das gestörte Problem lautet: $\forall z$ einer konvexen Nullumgebung $U \subset \boldsymbol{W}$ ist

$$\{k_1(v) - k_2(\boldsymbol{A}v - z) \mid v \in \boldsymbol{V}\} = \text{Inf}! \tag{15}$$

zu lösen. Die Störung drückt sich in einer Variation in den Ungleichungsrestriktionen in (12D) aus:

$$\boldsymbol{A}v \underset{K_W}{\geqq} h + z ,$$

wobei nach Umschreibung in die Form (7) und der Wahl von

$$U = \{(z_1, z_2)^T \mid \text{Max}\,(\|z_1\|, \|z_2\|) \leqq \varepsilon\}$$

die parametrische Aufgabe

$$\{\xi \mid \xi \in R^1, \xi \geqq 0, \xi - z_2 \geqq Ax - a \geqq z_1 - \xi, x \in \boldsymbol{X}\} = \text{Inf!} \tag{15'}$$

folgt. Für $z_1 = 0$ oder $z_2 = 0$ ist als Spezialfall die Störung von a selbst in der allgemeinen Störung enthalten. Gilt mit einer Konstanten $0 < C < \varepsilon$: $z_2 = -C$, $z_1 = -C$, so braucht die Nebenbedingung $|Ax - a| \leqq \xi$ nur gemäß $|Ax - a| \leqq \xi + C$ erfüllt zu sein. In allen Fällen würde sich der Infimalwert stabil verhalten.

4. Stabilität des Dualproblems

Liegt $a \in \boldsymbol{Y}$ nicht im Wertebereich $R(A)$ von A, so kann Annahme des Infimums $J = 0$ des Primalproblems nicht erfolgen. Das Dualproblem (13) oder (13') muß also für $a \in \boldsymbol{Y} \setminus R(A)$ instabil sein. Die einfache Gestalt von (13') ermöglicht, den Einfluß einer Störung $z^* \in U^* \subset \boldsymbol{V}^*$ in (12D), geschrieben in der zu (14) dualen Form, zu verfolgen. Das gestörte Dualproblem hat dann für jedes $z^* \in U^* \subset \boldsymbol{V}^*$ die Gestalt

$$\operatorname*{Sup}_{w^*} \{g^*(w^*) - f^*\,(\boldsymbol{A}^* w^* - z^*)\} = S(z^*) \tag{16}$$

mit

$$f^*\,(\boldsymbol{A}^* w^* - z^*) = \begin{cases} 0 & \text{für } \boldsymbol{A}^* w^* - z^* \leqq_{K_V^*} p \\ + \infty & \text{sonst}, \end{cases} \tag{16.1}$$

$$g^*(w^*) = \begin{cases} \langle w^*, h \rangle & \forall\, w^* \geqq_{K_W^*} 0 \\ - \infty & \text{sonst}. \end{cases} \tag{16.2}$$

Wir haben die Richtungsableitung an der Stelle $z^* = 0$

$$\lim_{\varepsilon \to +0} \frac{1}{\varepsilon} \left(S(\varepsilon z^*) - S(0)\right) \tag{17}$$

zu bilden $\forall z^* = (z_1^*, z_2^*) \in U^* \subset \boldsymbol{V}^* = \boldsymbol{X}^* \times R_1$. Im allein interessierenden Fall $S(z^*) \neq -\infty$ ist

$$\frac{1}{\varepsilon} S(\varepsilon z^*) = \operatorname*{Sup}_{w^*} \langle w^*, h \rangle\,, \tag{18}$$

wobei w^* den Bedingungen $w^* \geqq_{K_W^*} 0$ und

$$A^*\,(w_1^* - w_2^*) = \varepsilon z_1^*\,, \tag{19.1}$$

$$\|w_1^* - w_2^*\| \leqq 1 + \varepsilon z_2^* \tag{19.2}$$

zu genügen hat. Wegen $N(A^*) = \{0\}$ ist die Operatorgleichung (19.1) eindeutig oder nicht lösbar. Nur der lösbare Fall gibt endliches f^*. Sei $A^* (w_1^* - w_2^*) = z_1^*$ lösbar. Die Lösung heiße $\mathfrak{w}^*(z^*)$. Dann ist für hinreichend kleines ε wegen

$$A^* (\varepsilon w_1^* - \varepsilon w_2^*) = \varepsilon z_1^*$$

für $\varepsilon (w_1^* - w_2^*)$ sicher die Normierungsbedingung erfüllt, wenn $|z_2^*| < \frac{1}{2}$ gewählt wird. Folglich gilt der

Satz: *Der Infimalwert $J = 0$ des gegebenen Approximationsproblems wird genau dann angenommen, wenn für die $(z_1^*, z_2^*)^{\mathsf{T}} \in U^* = U_1^* \times U_2^* \subset \boldsymbol{X}^* \times R^1$ mit U_2^* hinreichend klein und für die z_1^*, für die die Operatorgleichung $A^* (w_1^* - w_2^*) = z_1^*$ lösbar ist, gilt*

$$\langle \mathfrak{w}^*(z^*), a \rangle \leqq C \tag{20}$$

mit einer Konstanten $C > 0$.

Im Instabilitätsfall müssen also bei „kleinen" z^* solche Schwankungen der Lösungen $\mathfrak{w}^*(z^*)$ möglich sein, daß das stetige lineare Funktional $\langle \mathfrak{w}^*, a \rangle$ über alle Schranken wächst. Dieser Sachverhalt erfährt eine interessante Erklärung, wenn wir zum Beispiel des elastischen Körpers in Abschnitt 2 zurückgehen (Kap. 5). Zuvor sei noch angemerkt, daß die Bedingung (20) unseres Satzes im Annahmefall erfüllt ist. Denn sei die konvexe Umgebung $U_1^* \subset \boldsymbol{X}^*$ durch

$$U_1^* = \{z_1^* \mid \sup \langle \mathring{x}, z_1^* \rangle \leqq 1\} \tag{21}$$

gegeben, wobei $\mathring{x}$ die Minimalstelle sei: $A\mathring{x} = a$. Dann ist

$$\begin{aligned} \langle \mathfrak{w}^*(z^*), a \rangle &= \langle \mathfrak{w}^*(z^*), A\mathring{x} \rangle \\ &= \langle A^* \mathfrak{w}^*(z^*), \mathring{x} \rangle \\ &= \langle z_1^*, \mathring{x} \rangle \leqq 1 \quad \forall z_1^* \in U_1^* , \end{aligned} \tag{22}$$

also (20) mit $C = 1$ erfüllt.

5. Analytische Ursache der Instabilität

Eine Lösung der 1. Randwertaufgabe der linearen Elastizitätstheorie (6) über Γ wird durch den GREENschen Tensor $G(P, Q)$ zu (6) gegeben ($G = G_{ik}$)

$$u(P) = \int_{\partial E} T_Q G(P, Q)\, x(Q)\, d\sigma_Q = Ax , \tag{23}$$

wobei der Vektordifferentialoperator T_Q in einem Punkt auf einem Flächenstück mit der (Außen-)Normalen n den Spannungsvektor bezüglich eines Verrückungsfeldes u erzeugt:

$$Tu = 2\mu \frac{\partial u}{\partial n} + \lambda n \operatorname{div} u + \mu (n \times \operatorname{rot} u) \tag{24}$$

und G die Bedingungen befriedigt, daß die Spalten bezüglich P die Gleichung $\Delta_P^* G = \delta_Q$ bei festem Q befriedigen, daß $G(P, Q) = (G(Q, P))^{\mathsf{T}}$ ist, daß die Spalten von

G bezüglich P die homogenen Randbedingungen befriedigen und daß schließlich die Singularität von G durch

$$\mathrm{Max}\ |G_{ik}| < \frac{C}{r} \quad \forall P, Q \in E \tag{25}$$

mit $r = \mathrm{dist}\ (P, Q)$ abgeschätzt ist.

Wir studieren nun die Operatorgleichung

$$A^* \ (w_1^* - w_2^*) = A^* \mathfrak{w}^* = z_1^* \tag{26}$$

für unsere Aufgabenstellung. Es ist dann $\boldsymbol{X}$ der BANACHraum der dreidimensionalen stetigen Vektorfunktionen über ∂E, $\boldsymbol{Y}$ entsprechend über $\Gamma \subset \mathrm{int}\ E$. A^* berechnet sich wie üblich aus

$$\langle \mathfrak{w}^*, Ax \rangle_{\boldsymbol{Y}} = \langle A^* \mathfrak{w}^*, x \rangle_{\boldsymbol{X}} , \tag{27}$$

und man erhält rechts ein lineares stetiges Funktional über $\boldsymbol{X}$. Entsprechend dem Darstellungssatz mit einem regulären Vektormaß μ wird es erzeugt durch

$$A^* \mathfrak{w}^* = \iint\limits_{\Gamma} T_Q G(P, Q)\ \mathrm{d}\mu_p . \tag{28}$$

Nun betrachten wir folgende Funktion für alle P

$$\iint\limits_{\Gamma} G(P, Q)\ \mathrm{d}\mu_p . \tag{29}$$

Sie genügt offenbar dem elliptischen Gleichungssystem (6) $\forall Q \in (\mathrm{int}\ E) \backslash \Gamma$, weil dies die Zeilen von G tun bezüglich Q. Weiter ist $\forall Q \in \partial E$ die Vektorfunktion (29) gleich Null, weil G die Randwerte Null befriedigt. Schließlich werde der Ableitungsoperator T_Q angewandt. Er kann wegen der Glattheit von G und $\Gamma \cap \partial E = \emptyset$ unter das Integral gezogen werden. Wegen (26) und (28) muß gerade sein

$$T_Q \iint\limits_{\Gamma} G(P, Q)\ \mathrm{d}\mu_p = z_1^* \quad \forall z_1^* \in U_1^* . \tag{30}$$

Was wir eben beschrieben haben heißt: die Vektorfunktion (29) erfüllt ein im HADAMARDschen Sinne inkorrekt gestelltes Problem für die Gleichung (6). Bei kleinen Änderungen der CAUCHYschen Anfangswerte, was durch (30) charakterisiert wird, verändern sich die Lösungen so, daß unsere Bedingung (20) bei geeignetem a verletzt werden kann.

Eben dieselbe Schlußweise kann auch bei L_2-Approximation und bei eindeutig lösbarem 1. Randwertproblem einer allgemeinen elliptischen Differentialgleichung oder bei einem Randanfangswertproblem mit DIRICHLETvorgaben für eine parabolische Differentialgleichung (man würde dieses Mal (bei $t = 0$) die Anfangswerte variieren und bei $t = T > 0$ auf einem Teilstück Γ des räumlichen Integrationsgebietes approximieren) angewendet werden [1], [6]. Jedesmal ergibt sich ein instabiles Randanfangswertproblem (im parabolischen Fall das backward-Problem für die zur gegebenen Gleichung adjungierte Gleichung), wodurch die Instabilität im entsprechenden Dualproblem verursacht werden kann.

Es ist noch zu zeigen, daß für den Operator A^* aus (28) gilt $N(A^*) = \{0\}$. Dies folgt, weil die Funktion (29) im Falle $z^*=0$ dem homogenen Cauchyschen Anfangswertproblem für $\Delta^* u=0$ genügt. Es genügt, die Randwerte x im Beispiel nur auf einer zusammenhängenden abgeschlossenen Teilmenge M von ∂E zu variieren, alle Schlüsse verlaufen dann ebenso, wenn $\boldsymbol{X}=C_0(M)\times\ldots\times C_0(M)$ gesetzt wird, wobei C_0 die über M stetigen Funktionen mit verschwindenden Randwerten sind, und x auf $\partial E\backslash M$ als identisch Null angesetzt wird. Da eine Lösung eindeutig bestimmt ist (Satz von Cauchy-Kowalewski), muß unter Berücksichtigung von (25) (29) identisch Null sein in $E\backslash\Gamma$. Mit der Schlußweise wie in [6] folgt daraus, daß auch $\mathfrak{w}^*$ selbst verschwindet. Wir denken uns nämlich eine beliebige stetige Vektorfunktion $g\in\boldsymbol{Y}$. Dann gibt es stets eine über E zweimal stetig differenzierbare Vektorfunktion u, die über ∂E verschwindet und für die längs Γ gilt

$$\|g-u\|<\varepsilon_1\,, \tag{31}$$

weil die C^2-Funktionen über Γ dicht liegen (in der Norm) in $\boldsymbol{Y}$. Dann sei $\Delta^* u=f$, und es gilt $\forall P\in E$ mit einer Konstanten $c\neq 0$

$$cu(P)=\iiint\limits_E G(P,Q)\,f(Q)\,\mathrm{d}v_Q\,. \tag{32}$$

Mit $\mathfrak{w}^*$, wofür μ das eineindeutig zugeordnete Vektormaß sei und wofür (29) verschwindet, folgt dann

$$c\langle\mathfrak{w}^*,u\rangle=\iint\limits_\Gamma\iiint\limits_E G(P,Q)\,f(Q)\,\mathrm{d}v_Q\,\mathrm{d}\mu_P\,. \tag{33}$$

Jetzt wird E unterteilt in eine Teilmenge E_1, die Γ ganz im Innern enthält, und $E\backslash E_1$. Aus (33) folgt

$$c\langle\mathfrak{w}^*,u\rangle=\iiint\limits_{E\backslash E_1} f(Q)\,\Big(\iint\limits_\Gamma G(P,Q)\,\mathrm{d}\mu_P\Big)\,\mathrm{d}\vartheta_Q + \iint\limits_\Gamma\iiint\limits_{E_1} G(P,Q)\,f(Q)\,\mathrm{d}v_Q\,\mathrm{d}\mu_P\,, \tag{34}$$

wobei der erste Summand rechts wegen

$$\iint\limits_\Gamma G(P,Q)\,\mathrm{d}\mu_P=0$$

verschwindet, der zweite wird wegen (25) beliebig klein, wenn nur das Lebesgue-Maß von E_1 hinreichend klein ist. Also verschwindet das Funktional $\mathfrak{w}^*$ wegen (31) auf einer in $\boldsymbol{Y}$ normdichten Menge, folglich stets.

Im Fall einer elliptischen Gleichung würde entsprechend der Unitätssatz von Cordes [2], bei der parabolischen Gleichung der von Friedman [3] herangezogen werden müssen.

Literatur

[1] Beckert, H.: Zur Steuerung der Stabilität in elastischen Körpern. ZAMM **52** (1972), 617–622.

[2] Cordes, H. O.: Über die eindeutige Bestimmtheit der Lösungen elliptischer Differentialgleichungen durch Anfangsvorgaben. Nachr. Akad. Wiss. Göttingen (1956), 239–258.

[3] Friedman, A.: Partial differential equations of parabolic type. Prentice Hall, Inc., Englewood Cliffs, N. Y. 1964.
[4] Göpfert, A.: Mathematische Optimierung in allgemeinen Vektorräumen. Teubner, Leipzig 1973.
[5] Göpfert, A.: Zur Anwendung der Dualitätstheorie in allgemeinen Räumen. Wiss. Zeitschr. der THLM **20** (1978) 2, 156–164.
[6] Göpfert, A.: Eine Anwendung des Unitätssatzes von Jto-Yamabe. Beitr. z. Anal. **1** (1971), 29–41.
[7] Joly, J. L., and P. J. Laurent: Stability and duality in convex minimisation problems. R.I.R.O. **5** (1971) 2, 3–42.
[8] Rockafellar, R. T.: Conjugate duality and optimization, Regional Conference Series in Appl. Math., Nr. 16, SIAM, Philadelphia 1974.

Summary

If the minimal value of the objective function is not attained for certain optimization problems of the approximation type then the corresponding dual problem is not stabil. This fact is based on the instability of the perturbated not correct stated problems with initial and boundary value conditions. The attainement of the minimal value is characterized by a condition which follows from the theory of stability of optimization problems.

Резюме

Если некоторые задачи программирования аппроксимационного типа не достигают своего минимального значения, тогда двойственная задача является неустойчивой. Этот факт находит своего объяснения в неустойчивости возмущенной, некорректно поставленной задачи с начальными и краевыми значениями. Достижение минимального значения характеризуется условием, вытекающим из теории устойчивости оптимизации.

Eingereicht bei der Redaktion: 2. 2. 1978; revidierte Fassung: 2. 5. 1978

Math. Operationsforsch. Statist., Ser. Optimization, Vol. 10 (1979) No. 1, 121–130

An Approximation Procedure for Stochastic Dynamic Programming based on Clustering of State and Action Spaces

V. Nollau [1] and A. Hahnewald-Busch [1]

Summary: In this note we investigate a time-discrete stochastic dynamic programming problem with countable state and action spaces. We introduce an approximation procedure for a numerical solution by decomposition of the state and also of the action space. The minimal value functions and the optimal policies of the Markovian Decision Processes constructed by clustering of both spaces are calculated by dynamic programming. Bounds for the minimal value functions will be obtained and convergence theorems are proved.

I. It is known by Hinderer [7], that a *stochastic dynamic program* is defined by a 5-typel (S, A, p, r, T). The *transition* of the system described by this program from a state $x \in S$ into any $y \in S$ under the influence of an action $u \in A$ happens in every of the N stages with the probability $p^u(x, y)$. Such a transition is connected with a *loss* $k^u(x, y)$, hence the *expected loss* for one stage is given by

$$r^u(x) = \sum_{y \in S} p^u(x, y)\, k^u(x, y) .$$

Furthermore, we assume to receive the *terminal loss* $T(z)$, if the system reaches the state z in the N-th stage. A sequence

$\mathfrak{d} = (d_0, d_1, \ldots, d_{N-1})$ of functions $d_k\colon x \to d_k(x) \in A \quad (x \in S;\ k = 0, 1, \ldots, N-1)$ is called a Markov*ian policy* and the set of all such policies is denoted by $\mathfrak{D}_N$.

A stochastic dynamic program, an initial distribution and a Markovian policy $\mathfrak{d}$ determine in a natural way a probability space $(\Omega, \mathfrak{F}, \mathsf{P}_\mathfrak{d})$ and so a Markovian chain $(X_0, X_1, \ldots, X_N)$ describing the considered decision process (cp. [4], [6]). Then the total loss is given by

$$R_N^\mathfrak{d} = \sum_{k=0}^{N-1} r^{d_k(X_k)}(X_k) + T(X_N) . \tag{1}$$

Based on the expected loss

$$V_0^\mathfrak{d}(x) = E_\mathfrak{d}\,(R_N^\mathfrak{d} \mid X_0 = x)$$

a policy $\mathfrak{d}^*$ is called an optimal policy, if

$$V_0^{\mathfrak{d}^*}(x) = \inf_{\mathfrak{d} \in \mathfrak{D}_N} V_0^\mathfrak{d}(x) \quad \text{for all} \quad x \in S$$

and $V_0^* = V_0^{\mathfrak{d}^*}$ is called the minimal value function.

[1] TU Dresden, Sektion Mathematik, DDR – 8027 Dresden, Mommsenstr. 13.

On condition that k^u and T are uniformly bounded by any C with $1 \leqq C < \infty$, the minimal value function V_0^* is the unique solution of the following equations (HINDERER [8]):

for $k=N$: $V_N^*(x)=T(x)$,

for $k=N-1, N-2, \ldots, 0$

$$V_k^u(x)=r^u(x)+\sum_{y\in S} p^u(x,y)\, V_{k+1}^*(y) \quad (u\in A) \tag{2}$$

and

$$V_k^*(x)=\inf_{u\in A} V_k^u(x)\,.$$

If S is finite with q elements we denote $V_k^*(x)$ by $V_k^*(x \mid q)$. It is well-known, that in practice it is rather difficult to calculate $V_0^*(x)$ for large sets of states and actions. For this reason we investigate approximative procedures based on clustering of the state and action spaces for calculation of $V_0^*(x)$. In the literature such procedures are given by KEMENY/SNELL [9] for generating MARKOV chains on a cluster state space and developed by BERTSEKAS [1] for the case when a random perturbation only exists for a finite subset of states. These algorithms differ essentially from the methods of HAHNEWALD-BUSCH/NOLLAU [5] and FOX [3] (based on choosing of finite original states in each level of approximation), which are generalized by the algorithms described here.

For this algorithm we consider two sequences $(q_m)_{m\in\mathbf{N}}$, $(z_n)_{n\in\mathbf{N}}$ of integers and $\mathfrak{S}^{(m)}$, $\mathfrak{A}^{(n)}$ are complete systems of q_m respectively z_n disjoint subsets of S respectively A. Furthermore, let π be a finite measure on $\mathfrak{P}(S)$[2] with $\pi(\{x\})>0$ $(x\in S)$. In the same way let σ be a finite measure on $\mathfrak{P}(A)$ with $\sigma(\{u\})>0$ $(u\in A)$. Based on the two sequences $(\mathfrak{S}^{(m)})_{m\in\mathbf{N}}$ and $(\mathfrak{A}^{(n)})_{n\in\mathbf{N}}$ we now define a family of dynamic programs (each of them accordingly to the original 5-tupel) by $(\mathfrak{S}^{(m)}, \mathfrak{A}^{(n)}, p_{m,n}, r_{m,n}, t_m)$:

1. $\mathfrak{S}^{(m)}$ is a finite set with q_m elements, the so-called cluster-state-space,
2. $\mathfrak{A}^{(n)}$ is a finite set with z_n elements, the so-called cluster-action-space,
3. $p_{m,n}$: $(X, Y, U)\to p_{m,n}^U(X, Y)$ is a function from $\mathfrak{S}^{(m)}\times\mathfrak{S}^{(m)}\times\mathfrak{A}^{(n)}$ into $[0, 1]$ given by

$$p_{m,n}^U(X,Y)=\sum_{u\in U}\frac{\sigma(\{u\})}{\sigma(U)}\sum_{x\in X}\frac{\pi(\{x\})}{\pi(X)}\sum_{y\in Y} p^u(x,y) \tag{3}$$

with $X, Y\in\mathfrak{S}^{(m)}$, $U\in\mathfrak{A}^{(n)}$.

It is easily seen, that (resulting from the definition) $p_{m,n}^U(X, Y)$ for every $U\in\mathfrak{A}^{(n)}$ is a MARKOVian transition probability with the properties

$$p_{m,n}^U(X,Y)\geqq 0 \quad\text{and}\quad \sum_{Y\in\mathfrak{S}^{(m)}} p_{m,n}^U(X,Y)=1.$$

4. $r_{m,n}$: $(X, U)\to r_{m,n}^U(X)=\sum_{Y\in\mathfrak{S}^{(m)}} p_{m,n}^U(X,Y)\, k_{m,n}^U(X,Y)$ with

$$k_{m,n}^U(X,Y)=\sum_{u\in U}\frac{\sigma(\{u\})}{\sigma(U)}\sum_{x\in X}\frac{\pi(\{x\})}{\pi(X^u)}\sum_{y\in Y}\frac{p^u(x,y)\,k^u(x,y)}{p_{m,n}^{\{u\}}(\{x\},Y)} \tag{4}$$

[2] $\mathfrak{P}(S)$ is the set of all subsets of S.

for $X, Y \in \mathfrak{S}^{(m)}, U \in \mathfrak{A}^{(n)}$, and

$$X^u = \{x \in X : p^{\{u\}}_{m,n}(\{x\}, Y) > 0\}.$$ [3]

Obviously $k^U_{m,n}(X, Y)$ and in consequence $r^U_{m,n}(X)$ are bounded by C,

5. $t_m: X \to t_m(X)$, defined on $\mathfrak{S}^{(m)}$, is the non-negative terminal loss function from $\mathfrak{S}^{(m)}$ into $[0, C]$, given by

$$t_m(X) = \frac{1}{\pi(X)} \sum_{x \in X} \pi(\{x\})\, T(x) \quad (X \in \mathfrak{S}^{(m)}).$$

In the same way as before we define a system of functions $\mathfrak{B}^*_k$ as follows

for $k = N$: $\mathfrak{B}^*_N(X; m, n) = t_m(X)$ $(X \in \mathfrak{S}^{(m)})$,

for $k = N-1, N-2, \ldots, 0$ recursively

$$\mathfrak{B}^U_k(X; m, n) = \sum_{Y \in \mathfrak{S}^{(m)}} p^U_{m,n}(X, Y)\, [k^U_{m,n}(X, Y) + \mathfrak{B}^*_{k+1}(Y; m, n)]$$

and

$$\mathfrak{B}^*_k(X; m, n) \equiv \min_{U \in \mathfrak{A}^{(n)}} \mathfrak{B}^u_k(X; m, n)\,. \tag{5}$$

Thus the function $\mathfrak{B}^*_k(\cdot; m, n)$ is the unique solution of the system (5). Simultaneously we receive the optimal decision $d^*_k(X; m, n)$ for every $X \in \mathfrak{S}^{(m)}$ and $k = N-1, N-2, \ldots, 0$.

II. Now we can solve every of these new stochastic dynamic programs, because the cluster-state-spaces $\mathfrak{S}^{(m)}$ and $\mathfrak{A}^{(n)}$ are finite. The computational algorithms are rather simple, but it is interesting to examine the convergence properties when m and n tend to infinity. At first we introduce the concept of admissibility for the decomposition of the state and action spaces in the following way:

Definition 1: Let $(\delta_m)_{m \in N}$ and $(\varepsilon_n)_{n \in N}$ be two strictly decreasing sequences of positive numbers which tend to zero. Then a two parameter family $(\mathfrak{S}^{(m)}, \mathfrak{A}^{(n)})_{m,n \in N}$ of decompositions of (S, A) is said to be *admissible* (with respect to (δ_m) and (ε_n)), if

(i) for every $x \in X$ and $X \in \mathfrak{S}^{(m)}$ holds

$$\pi(X) - \pi(\{x\}) \leq \delta_m\,,$$

(ii) for every $(x, y) \in S \times S$ and $U \in \mathfrak{A}^{(n)}$ holds

$$\left| \frac{1}{\sigma(U)} \sum_{u \in U} \sigma(\{u\})\, p^u(x, y) - p^u(x, y) \right| + \left| \frac{1}{\sigma(U)} \sum_{u \in U} \sigma(\{u\})\, k^u(x, y) - k^u(x, y) \right| \leq \varepsilon_n\,.$$

Furthermore, we shall use the following convergence type introduced by Carathéodory [2].

Definition 2: Let $(f_m)_{m \in N}$ be a sequence of realvalued functions defined on $\mathfrak{P}(B)$ and $\mathfrak{B}^{(m)}$ a complete system of disjoint subsets of B. Then we shall say,

[3] If $\pi(X^u) = 0$, we define $k^U_{m,n}(X, Y) = 0$.

that $(f_m)_{m\in N}$ *converges continuously* to a realvalued function f defined on B, if for every sequence (b_m) with $b_m \in \mathfrak{B}^{(m)}$ tending to $b \in B$[4] the sequence $(f_m(b_m))_{m\in N}$ converges to $f(b)$ for $m \to \infty$.

Based on this definition we obtain the following

Lemma 1: *Let* $(\mathfrak{S}^{(m)}, \mathfrak{A}^{(n)})_{m,n\in N}$ *be an admissible two parameter family of decompositions of* (S, A). *Then for every state* $x_0 \in S$ *there exists an integer* m_0 *with the property, that* $\{x_0\} \in \mathfrak{S}^{(m)}$ *for* $m \geqq m_0$.

Proof: We denote $\pi(\{x_0\}) = \alpha$, whereby α is a positive number. Then we get for $x_0 \in X_m$ $(X_m \in \mathfrak{S}^{(m)})$

$$\pi(X_m) = \sum_{x \in X_m} \pi(\{x\}) \geqq \alpha .$$

Regarding the admissibility of $(\mathfrak{S}^{(m)}, \mathfrak{A}^{(n)})_{m,n\in N}$ there exists an integer $m_0 = m_0(\alpha)$ with the property

$$\pi(X_m) - \pi(\{x\}) \leqq \delta_{m_0} < \frac{\alpha}{2} \qquad (*)$$

for every $x \in X_m$ and all $X_m \in \mathfrak{S}^{(m)}$ $(m \geqq m_0)$.

Now let us assume the existence of an integer m with $m \geqq m_0$ such that $x_0 \in X_m$ but $\{x_0\} \neq X$ holds.

Then it follows $\sum_{\substack{x \in X_m \\ x \neq x_0}} \pi(\{x\}) < \frac{\alpha}{2}$

and in particular $\pi(\{x\}) < \frac{\alpha}{2}$ for every $x \in X_m \setminus \{x_0\}$.

Hence, for every $x \in X_m \setminus \{x_0\}$ we have

$$\pi(X_m) - \pi(\{x\}) > \frac{\alpha}{2} .$$

But this is impossible by (*) and therefore $\{x_0\} = X_m$ holds. ∎

Now we construct common upper and lower bounds for the functions $V_k^*(x)$ and $V_k^*(X; m, n)$, which describe the expected total loss on the k-th stage for the original optimization problem and the optimization problem on the clustered spaces $(\mathfrak{S}^{(m)}, \mathfrak{A}^{(n)})$. For this purpose let be

$$\underline{p}_{m,n}^U(X, Y) = \inf_{(u,x,y)\in(U,X,Y)} p^u(x, y) ,$$
$$\bar{p}_{m,n}^U(X, Y) = \sup_{(u,x)\in(U,X)} \sum_{y\in Y} p^u(x, y) ,$$
$$\underline{k}_{m,n}^U(X, Y) = \inf_{(u,x,y)\in(U,X,Y)} k^u(x, y) ,$$
$$\bar{k}_{m,n}^U(X, Y) = \sup_{(u,x,y)\in(U,X,Y)} k^u(x, y)$$

with $(X, Y \in \mathfrak{S}^{(m)}, U \in \mathfrak{A}^{(n)})$.

---

[4] That is $\overline{\lim_{m\to\infty}}\, b_m = \underline{\lim_{m\to\infty}}\, b_m = \{b\}$.

Furthermore we are considering the following terms

$$\underline{L}_N^*(X; m) = \inf_{x \in X} T(x), \quad \bar{L}_N^*(X; m) = \sup_{x \in X} T(x)\,^5, \tag{5}$$

$$\underline{L}_k^U(X; m, n) = \sum_{Y \in \mathfrak{S}^{(m)}} \underline{p}_{m,n}^U(X, Y)\,(\underline{k}_{m,n}^U(X, Y) + \underline{L}_{k+1}^*(Y; m, n))$$

with

$$\underline{L}_k^*(X; m, n) = \min_{U \in \mathfrak{A}^{(n)}} \underline{L}_k^U(X; m, n)$$

and

$$\bar{L}_k^U(X; m, n) = \sum_{Y \in \mathfrak{S}^{(m)}} \bar{p}_{m,n}^U(X, Y)\,(\bar{k}_{m,n}^U(X, Y) + \bar{L}_{k+1}^*(Y; m, n))$$

with

$$\bar{L}_k^*(X; m, n) = \min_{U \in \mathfrak{A}^{(n)}} \bar{L}_k^*(X; m, n)$$

for $k = N-1, N-2, \ldots, 0$.

The following inequalities are obvious:

$$\underline{L}_k^*(X; m, n) \leqq \mathfrak{V}_k^*(X; m, n) \leqq \bar{L}_k^*(X; m, n)$$

and

$$\underline{L}_k^*(X; m, n) \leqq V_k^*(x) \leqq \bar{L}_k^*(X; m, n) \quad (x \in X)$$

for $k = N, N-1, N-2, \ldots, 0$, $X \in \mathfrak{S}^{(m)}$ and $m, n \in \boldsymbol{N}$.
As an immediate consequence of theorem 1 and theorem 2 in [5], proved in this note, we obtain the following

Theorem 1: *Let (S, A, p, r, T) be a stochastic dynamic program with a finite set A of actions. Furthermore let $(\mathfrak{S}^{(m)}, \mathfrak{A}^{(n)})_{m,n \in \boldsymbol{N}}$ be a family of admissible decompositions of (S, A) with $\mathfrak{A}^{(n)} = A$ for all $n \in \boldsymbol{N}$. Then for all $k = 0, 1, \ldots, N$*

$$\underline{L}_k^*(\cdot; m, n), \quad \bar{L}_k^*(\cdot; m, n) \quad \text{and} \quad \mathfrak{V}_k^*(\cdot; m, n)$$

converge continuously to $V_k^*(\cdot)$ if m tends to infinity.[6]

In connection with this proposition we obtain moreover the following important result (cp. [5]):

There is a finite integer m depending on the considered state, such that the optimal decision with respect to the optimization problem on the cluster-space $(\mathfrak{S}^{(m)}, A)$ is also an optimal decision of the original problem.

Now we consider a family of admissible decompositions

$$(\mathfrak{S}^{(m)}, \mathfrak{A}^{(n)})_{m,n \in \boldsymbol{N}} \quad \text{of} \quad (S, A) \quad \text{with} \quad \mathfrak{S}^{(m)} \equiv S \quad \text{for all} \quad m \in \boldsymbol{N}.$$

Then we get at first the

---

[5] We have $\underline{L}_N^*(X; m, n) = \underline{L}_N^*(X; m)$ and $\bar{L}_N^*(X; m, n) = \bar{L}_N^*(X; m)$, because $\underline{L}_N^*(X; m)$ and $\bar{L}_N^*(X; m)$ are independent of n.

[6] Let us remark, that $\underline{L}_k^*(\cdot; m, n)$, $\bar{L}_k^*(\cdot; m, n)$ and $\mathfrak{V}_k^*(\cdot; m, n)$ are independent of n, because $\mathfrak{A}^{(n)} \equiv A$ for all $n \in \boldsymbol{N}$.

Lemma 2: *Let be S a finite state space and $(\mathfrak{S}^{(m)}, \mathfrak{A}^{(n)})_{m,n\in N}$ an admissible family of decompositions of (S, A) with $\mathfrak{S}^{(m)} \equiv S$ for all $m \in N$.*
Then for any sequence $(U_n)_{n\in N}$ with $U_n \in \mathfrak{A}^{(n)}$

(i) $$\lim_{n\to\infty} (\bar{p}_{m,n}^{U_n}(\{x\}, \{y\}) - \underline{p}_{m,n}^{U_n}(\{x\}, \{y\}))$$
$$= \lim_{n\to\infty} (\bar{p}_n^{U_n}(\{x\}, \{y\}) - \underline{p}_n^{U_n}(\{x\}, \{y\})) = 0 \quad (x, y \in S)$$

and

(ii) $$\lim_{n\to\infty} (\bar{k}_{m,n}^{U_n}(\{x\}, \{y\}) - \underline{k}_{m,n}^{U_n}(\{x\}, \{y\}))$$
$$= \lim_{n\to\infty} (\bar{k}_n^{U_n}(\{x\}, \{y\}) - \underline{k}_n^{U_n}(\{x\}, \{y\})) = 0 \quad (x, y \in S).$$

Proof: Let ε be any positive number. Then by the admissibility of $(S, \mathfrak{A}^{(n)})_{n\in N}$ there exists an integer $n_0 = n_0(\varepsilon)$ with $\varepsilon_{n_0} < \frac{\varepsilon}{4}$, such that for all $u \in U_n$ $(n \geqq n_0)$ holds

$$\left| \frac{1}{\sigma(U_n)} \sum_{u\in U_n} \sigma(\{u\})\, p^u(x, y) - p^u(x, y) \right| < \varepsilon_n < \frac{\varepsilon}{4}. \qquad (*)$$

Furthermore, by definition of $\bar{p}^{U_n}$ and $\underline{p}^{U_n}$ there exist two actions $\tilde{u}, \tilde{\tilde{u}} \in U_n$ with

$$|\bar{p}_n^{U_n}(\{x\}, \{y\}) - p^{\tilde{u}}(x, y)| + |\underline{p}_n^{U_n}(\{x\}, \{y\}) - p^{\tilde{\tilde{u}}}(x, y)| < 2\varepsilon_n. \qquad ({*\atop *})$$

By (*) and $({*\atop *})$ we obtain the following relation

$$|\bar{p}_n^{U_n}(\{x\}, \{y\}) - \underline{p}_n^{U_n}(\{x\}, \{y\})|$$
$$\leqq |\bar{p}_n^{U_n}(\{x\}, \{y\}) - p^{\tilde{u}}(x, y)| + \left| p^{\tilde{u}}(x, y) - \frac{1}{\sigma(U_n)} \sum_{u\in U_n} \sigma(\{u\})\, p^u(x, y) \right|$$
$$+ \left| \frac{1}{\sigma(U_n)} \sum_{u\in U_n} \sigma(\{u\})\, p^u(x, y) - p^{\tilde{\tilde{u}}}(x, y) \right| + |p^{\tilde{\tilde{u}}}(x, y) - \underline{p}_n^{U_n}(\{x\}, \{y\})|$$
$$< 4\varepsilon_n < \varepsilon \quad \text{for all} \quad n \geqq n_0(\varepsilon).$$

Thus the proposition (i) is proved. Part (ii) of this lemma may be proved in analogous manner. ▌

Theorem 2: *Let (S, A, p, r, T) be a stochastic dynamic program with a finite set S of q states. Furthermore, let $(\mathfrak{S}^{(m)}, \mathfrak{A}^{(n)})_{m,n\in N}$ be a family of admissible decompositions of (S, A) with $\mathfrak{S}^{(m)} \equiv S$ for all $m \in N$.*
Then for all $k = 0, 1, \ldots, N$

$$\underline{L}_k^*(\{x\}; m, n) = \underline{L}_k^*(\{x\}; n),$$
$$\bar{L}_k^*(\{x\}; m, n) = \bar{L}_k^*(\{x\}; n) \quad \textit{and}$$
$$\mathfrak{B}_k^*(\{x\}; m, n) = \mathfrak{B}_k^*(\{x\}; n) \quad (x \in S)$$

converge to $V_k^*(x \mid q)$ for $n \to \infty$.

Proof: We shall prove these properties recursively in the following way: At first let us consider

$k=N$: Then for every $x \in S$ we have

$$\bar{L}_N^*(\{x\};n)=t_n(\{x\})=\underline{L}_N^*(\{x\};n).$$

Further we obtain for the stage
$k=N-1$:

$$\begin{aligned}
&\bar{L}_{N-1}^*(\{x\};n)-\underline{L}_{N-1}^*(\{x\};n)\\
&=\min_{U\in\mathfrak{A}^{(n)}}\sum_{y\in S}\bar{p}_n^U(\{x\},\{y\})\,(\bar{k}_n^U(\{x\},\{y\})+\bar{L}_N^*(\{y\};n))\\
&\quad-\min_{U\in\mathfrak{A}^{(n)}}\sum_{y\in S}\underline{p}_n^U(\{x\},\{y\})\,(\underline{k}_n^U(\{x\},\{y\})+\underline{L}_N^*(\{y\};n)).
\end{aligned}$$

Let be $U_n^*\in\{U:\underline{L}_{N-1}^U(\{x\};n)=\underline{L}_{N-1}^*(\{x\};n)\}$, then we obtain for any U_n^*

$$\bar{L}_{N-1}^*(\{x\};n)=\min_{U\in\mathfrak{A}^{(n)}}\bar{L}_{N-1}^U(\{x\};n)\leqq\bar{L}_{N-1}^{U_n^*}(\{x\};n)$$

and

$$\begin{aligned}
&\bar{L}_{N-1}^*(\{x\};n)-\underline{L}_{N-1}^*(\{x\};n)\\
&\leqq\bar{L}_{N-1}^{U_n^*}(\{x\};n)-\underline{L}_{N-1}^{U_n^*}(\{x\};n)\\
&=\sum_{y\in S}\bar{p}_n^{U_n^*}(\{x\},\{y\})\,(\bar{k}_n^{U_n^*}(\{x\},\{y\})+\bar{L}_N^*(\{y\};n))\\
&\quad-\sum_{y\in S}\underline{p}_n^{U_n^*}(\{x\},\{y\})\,(\underline{k}_n^{U_n^*}(\{x\},\{y\})+\underline{L}_N^*(\{y\};n))\\
&=\sum_{y\in S}\bar{p}_n^{U_n^*}(\{x\},\{y\})\,(\bar{k}_n^{U_n^*}(\{x\},\{y\})-\underline{k}_n^{U_n^*}(\{x\},\{y\}))\\
&\quad+\sum_{y\in S}\underline{k}_n^{U_n^*}(\{x\},\{y\})\,(\bar{p}_n^{U_n^*}(\{x\},\{y\})-\underline{p}_n^{U_n^*}(\{x\},\{y\}))\\
&\quad+\sum_{y\in S}t_n(\{y\})\,(\bar{p}_n^{U_n^*}(\{x\},\{y\})-\underline{p}_n^{U_n^*}(\{x\},\{y\})).
\end{aligned}$$

For any $\varepsilon>0$ let $n_0=n_0(\varepsilon)$ be an integer with $\varepsilon_n<\frac{\varepsilon}{12qC}$ for $n\geqq n_0$.

As $\bar{p}_n^{U_n^*}(\{x\},\{y\})$, $\underline{k}_n^{U_n^*}(\{x\},\{y\})$ and $t_n(\{y\})$ are bounded by C, we obtain by the admissibility of $(S,\mathfrak{A}^{(n)})_{n\in\mathbf{N}}$ and lemma 2

$$\bar{L}_{N-1}^*(\{x\};n)-\underline{L}_{N-1}^*(\{x\};n)\leqq 3qC\cdot 4\varepsilon_n<\varepsilon\quad(n\geqq n_0)$$

and consequently

$$\lim_{n\to\infty}(\bar{L}_{N-1}^*(\{x\};n)-\underline{L}_{N-1}^*(\{x\};n))=0\,.$$

By the relations

$$\underline{L}_{N-1}^*(\{x\};n)\leqq\mathfrak{V}_{N-1}^*(\{x\};n)\leqq\bar{L}_{N-1}^*(\{x\};n)$$

and

$$\underline{L}_{N-1}^*(\{x\};n)\leqq V_{N-1}^*(x\mid q)\leqq\bar{L}_{N-1}^*(\{x\};n)$$

we obtain the proposition for the $(N-1)$-th stage and in analoguous manner recursively for all $k=N-2, N-3, \ldots, 0$. ▮

III. Summarizing these results of the second part of this note we shall obtain a general theorem concerning an approximation procedure for countable state and action spaces.

Theorem 3: *Let (S, A, p, r, T) be a stochastic dynamic program with a family $(\mathfrak{S}^{(m)}, \mathfrak{A}^{(n)})$ $m, n \in \boldsymbol{N}$ of admissible decompositions of (S, A) and $m = m(n)$ any sequence of integers with the properties*

$$\lim_{n \to \infty} q_{m(n)} \cdot \varepsilon_n = 0 \quad \textit{and} \quad \lim_{n \to \infty} m(n) = \infty .$$ [7]

Then the functions $\mathfrak{V}_k^*(\cdot\,; m, n) = \mathfrak{V}_k^*(\cdot\,; m(n), n)$ for all $k = 0, \ldots, N$ converge continuously to $V_k^*(\cdot)$ for $n \to \infty$.

Proof: By lemma 1 for any $x \in S$ there exists an integer $m_0 = m_0(x)$ with the property, that $\{x\} \in \mathfrak{S}^{(m)}$ for all $m \geqq m_0$. Further let be $(X_m)_{m \in \boldsymbol{N}}$ with $X_m \in \mathfrak{S}^{(m)}$ converging to $\{x\}$. Then there exists an integer $m' = m'(x)$, so that $x \in X_m$ for all $m \geqq m'$. Consequently for all $m \geqq m_0' = \max(m_0, m')$ the sequence $(X_m)_{m \in \boldsymbol{N}}$ is stationary, i.e. $X_m = \{x\}$ for $m \geqq m_0'$. Further we denote

$$\underline{L}_k^*(\{x\}; n) \quad \text{and} \quad \bar{L}_k^*(\{x\}; n) \quad \text{by}$$
$$\underline{L}_k^*(\{x\}; n \mid q) \quad \text{and} \quad \bar{L}_k^*(\{x\}; n \mid q) \quad \text{respectively,}$$

if the state space consists of q elements.
The proof of theorem 2 yields, that

$$\lim_{n \to \infty} (\bar{L}_k^*(\{x\}; n) - \underline{L}_k^*(\{x\}; n)) = 0$$

holds for a fixed number q of states and so for $q = q_{m(n)}$ with the property we suppose above. In this way we obtain the result

$$\lim_{n \to \infty} (\bar{L}_k^*(\{x\}; n \mid q_{m(n)}) - \underline{L}_k^*(\{x\}; n \mid q_{m(n)})) = 0 .$$

Now we are looking at the boundaries $\bar{L}_k^*$ and $\underline{L}_k^*$ and are remarking, that

$$\underline{L}_k^*(\{x\}; n \mid q_{m(n)}) = \underline{L}_k^*(\{x\}; m(n), n) ,$$
$$\bar{L}_k^*(\{x\}; n \mid q_{m(n)}) = \bar{L}_k^*(\{x\}; m(n), n)$$

for all $n \geqq n_0$ with $m(n_0) \geqq m_0$.

By the relations

$$\underline{L}_k^*(\{x\}; m(n), n) \leqq \mathfrak{V}_k^*(X; m(n), n) \leqq \bar{L}_k^*(\{x\}; m(n), n) \quad (x \in X)$$
$$\underline{L}_k^*(\{x\}; m(n), n) \leqq V_k^*(x) \leqq \bar{L}_k^*(\{x\}; m(n), n)$$

we now obtain the continuous convergence of

$$\mathfrak{V}_k^*(\cdot\,; m(n), n) \quad \text{to} \quad V_k^* . \blacksquare$$

Remark: For a numerical approximation of the value $V_k^*(x)$ for every $x \in S$ and $k = 0, 1, \ldots, N$ the measures π and σ may be chosen in form of any weight function (usually dependent on the "importance" of every state $x \in S$ and every

---

[7] q_m is the number of elements of $\mathfrak{S}^{(m)}$ as defined before.

action $u \in A$). By defining the sequences $(\delta_m)_{m \in N}$ and $(\varepsilon_n)_{n \in N}$ for any $\varepsilon > 0$ an integer $n_0(\varepsilon)$ and so an integer $m_0 = m_0(n_0)$ exists with $\{x\} \in \mathfrak{S}^{(m_0)}$ and

$$|\mathfrak{V}_k^*(\{x\}; m, n) - V_k^*(x)| < \varepsilon \quad \text{for all} \quad m \geqq m_0 \quad \text{and} \quad n \geqq n_0 .$$

Especially the different speeds of decomposition in the state and action spaces will be influenced by the convergence behavior of the sequence $(q_{m(n)} \cdot \varepsilon_n)_{n \in N}$. Moreover, the order of convergence of the upper and lower bounds for the value functions is determined by the order of convergence of $(q_{m(n)} \cdot \varepsilon_n)_{n \in N}$.

References

[1] BERTSEKAS, D.: Convergence of Discretization Procedures in Dynamic Programming. IEEE Autom. Control. **20** (1975), 415–419.

[2] CARATHEODORY, C.: Funktionentheorie I. Basel 1950.

[3] FOX, B. L.: Discretizing Dynamic Programs. Journ. Optim. Theory Appl. **11** (1973) 3.

[4] GIRLICH, H.-J.: Diskrete stochastische Entscheidungsprozesse. Leipzig 1973.

[5] HAHNEWALD-BUSCH, A., und V. NOLLAU: An Approximation Procedure for Stochastic Dynamic Programming in Countable State Space. Math. Operationsforsch. Statist., Ser. Optimization **9** (1978) 1, 109–117.

[6] HINDERER, K.: Foundations of non-stationary dynamic programming. Lecture Notes in Oper. Res. and Math. Syst. No. 33, Berlin-Heidelberg-New York 1970.

[7] HINDERER, K.: Neuere Resultate der Stochastischen Dynamischen Optimierungs Zeitschr. f. angew. Math. und Mech. **55** (1975), T16–T26.

[8] HINDERER, K.: Does the value iteration for finite-stage discrete dynamic programs holds "always"? Bonner Math. Schriften **98** (1977), 41–56.

[9] KEMENY, J. G., und J. L. SNELL: Finite Markov Chains. Moskau 1970 (russ. Übersetzung).

[10] WHITT, W.: Approximation of Dynamic Programs (to appear in Math. Oper. Res. **3** (1978)).

Zusammenfassung

In dieser Arbeit wird ein zeitdiskretes stochastisches dynamisches Optimierungsproblem mit abzählbarem Zustands- und Entscheidungsraum untersucht. Zur numerischen Lösung wird eine Näherungsmethode mit Hilfe von Zerlegungen sowohl des Zustands- als auch des Entscheidungsraumes angegeben. Die minimalen Wertfunktionen und die optimalen Politiken des MARKOVschen Entscheidungsprozesses, der durch Vergröberung beider Räume konstruiert wird, werden mittels der Methode der dynamischen Optimierung berechnet. Für die minimalen Wertfunktionen werden Schranken angegeben und Konvergenzsätze bewiesen.

Резюме

Исследуется проблема дискретного по времени стохастического динамического программирования со счетными пространствами состояний и управлений. Излагается приближенная процедура для вычислительного решения с помощью разложения пространств состояний и управлений. Вычисляются с помощью динамического программирования оптимальные функции и оптимальные политики управления процессом решения Марковского типа, сконструированным с помощью последовательности конечных подмножеств этих пространств. Выводятся грани оптимальных значений функций и доказывается сходимость метода.

Eingereicht bei der Redaktion: 8. 12. 1977

Math. Operationsforsch. Statist., Ser. Optimization, Vol. 10 (1979) No. 1, 131–142

Anwendung der Dualitätstheorie auf Probleme der optimalen Steuerung[1]

H. BENKER[2] und H. DIETRICH[2]

Zusammenfassung: Es werden normminimale und quadratische Probleme der optimalen Steuerung im HILBERTraum betrachtet. Für diese Aufgaben werden Optimalitätsbedingungen angegeben und Dualitätssätze aufgestellt. Die erhaltenen Dualitätsaussagen werden dazu verwendet, untere Schranken für die betrachteten Funktionale zu berechnen.

1. Einleitung

Wir betrachten im HILBERTraum die folgenden beiden Aufgaben der optimalen Steuerung

$$J_1(u) = j_1(Q, u) = \|Q - R\|_1 + K\,\|u\|_2 \overset{!}{=} \underset{(Q,u)\in V_i}{\text{Minimum}} \tag{1}$$

und

$$J_2(u) = j_2(Q, u) = \|Q - R\|_1^2 + K\,\|u\|_2^2 \overset{!}{=} \underset{(Q,u)\in V_i}{\text{Minimum}} \tag{2}$$

mit

$\|\ldots\|_i$ – Norm im HILBERTraum H_i
K – Konstante $\geqq 0$
R – gegebenes Element $\in H_1$.

Als Vergleichsmengen V_i verwenden wir dabei:

$$V_1 = \{(Q, u) \mid AQ + Bu + f = 0;\, Q \in H_1,\, u \in U \subset H_2\}$$
$$V_2 = \{(Q, u) \mid Q = Lu + f;\, Q \in H_1,\, u \in U \subset H_2\}$$

mit $f \in H_1$ bzw. H_3 (gegebenes Element) und $A \in L\,(H_1 \to H_3)$, $B \in L\,(H_2 \to H_3)$, $L \in L\,(H_2 \to H_1)$, d. h., A, B und L sind linear. Zusätzlich seien B und L beschränkt, und es existiere A^{-1} (ebenfalls beschränkt). Weiterhin liege der Definitionsbereich von A dicht in H_1, so daß der adjungierte Operator A^{T} existiert. Im Falle der speziellen Beschränkung $\|u\|_2 \leqq 1$, d. h.

$$U = \{u \in H_2 \mid \|u\|_2 \leqq 1\},$$

bezeichnen wir die entsprechende Vergleichsmenge V_i mit V_i^1.

---

[1] Vortrag, gehalten auf der Tagung „Mathematische Optimierung – Theorie und Anwendungen“ in Eisenach (DDR), 5.–10. Dezember 1977.

[2] Technische Hochschule „Carl Schorlemmer“ Leuna-Merseburg, Sektion Mathematik und Rechentechnik, DDR – 42 Merseburg, Geusaer Str.

Auf Grund unserer Voraussetzungen sind also nur lineare Zustandsgleichungen der Form $AQ+Bu+f=0$ und $Q=Lu+f$ zugelassen. Weiterhin nehmen wir immer an, daß die Steuermenge U beschränkt, konvex und abgeschlossen ist, so daß auf Grund des WEIERSTRASSschen Satzes die Existenz einer Lösung für die Aufgaben (1) und (2) gesichert ist. Die eindeutige Lösbarkeit der Zustandsgleichungen für alle $u \in U$ setzen wir ebenfalls voraus, so daß die Schreibweise $J_1(u)$ und $J_2(u)$ in (1) bzw. (2) gerechtfertigt ist.[3]

Mit den Aufgaben (1) und (2) kann eine Vielzahl von konkreten Problemen der optimalen Steuerung in Funktionenräumen behandelt werden, da

– die Funktionale aus (1) und (2) praktisch häufig vorkommen: Ein vorgegebener Zustand R ist „möglichst gut" anzunähern, wobei im Falle $K>0$ der „Steuerungsaufwand" mit zu berücksichtigen ist;

– mit der linearen Operatorgleichung $AQ+Bu+f=0$ lineare Differential-, Integral- bzw. Integrodifferentialgleichungen beschrieben werden können (wenn wir diese Gleichungen z. B. mittels GREENscher oder RIEMANNscher Funktionen lösen, erhalten wir eine Gleichung der Form $Q=Lu+f$, die wir in V_2 verwenden).

Besonders vorteilhaft erweist sich die hier betrachtete Aufgabenstellung bei Aufgaben, die durch partielle Differentialgleichungen beschrieben werden, da sich hier die einzelnen Probleme i. allg. nur durch den Typ der Differentialgleichung unterscheiden. Wenn wir die Vergleichsmenge V_2 benutzen, schreiben wir die Aufgaben (1) und (2) immer in der Form

$$J_1^L(u)=\|Lu+r\|_1+K\,\|u\|_2 \overset{!}{=} \underset{u \in U}{\text{Minimum}} \tag{1'}$$

bzw.

$$J_2^L(u)=\|Lu+r\|_1^2+K\,\|u\|_2^2 \overset{!}{=} \underset{u \in U}{\text{Minimum}} \tag{2'}$$

mit

$$r=f-R\,.$$

Da beide Aufgaben für $K=0$ äquivalent sind, gelten folglich auch in diesem Fall die für eine der Aufgaben angegebenen Eigenschaften ebenfalls für die andere. Im folgenden betrachten wir aber hauptsächlich den Fall $K>0$, für den beide Aufgaben i. allg. verschiedene Lösungen liefern.

Der Schwerpunkt der folgenden Ausführungen liegt darauf, durch Aufstellung dualer Aufgaben die Struktur des Problems zu vereinfachen (wenn möglich!!) und Schranken für die Funktionale zu berechnen. Zur Dualisierung verwenden wir dabei die bekannten Theorien nach FENCHEL und LAGRANGE. In einem vorangehenden Abschnitt werden noch kurz Optimalitätsbedingungen angegeben und einige weitere Eigenschaften bewiesen, die für die folgenden Betrachtungen nützlich sind.

---

[3] $Q(u)$ bedeute dabei im weiteren die zu $u \in U$ gehörige Lösung der entsprechenden Zustandsgleichung.

Während es für Steuerungsprobleme mit konzentrierten Parametern schon eine Reihe von Arbeiten zur Dualitätstheorie gibt (siehe z. B. [2, 3, 4, 7, 18]), wurden Aufgaben mit verteilten Parametern und abstrakte Aufgaben der Form (1), (2) in bezug auf Dualitätsaussagen bisher wenig betrachtet. Hier sind uns nur die Arbeiten [5, 6, 9, 10, 21] bekannt.

2. Eigenschaften

Wie wir im Abschnitt 1. bereits bemerkt haben, ist die Existenz mindestens einer Lösung unter unseren gegebenen Voraussetzungen für beide Aufgaben stets gesichert. Da das Funktional aus (2) für $K > 0$ stark konvex ist, existiert hier sogar nur genau eine Lösung. Kommen wir nun zu Optimalitätsbedingungen für die Aufgaben (1) und (2).

Satz 1: *Ein Element* $(Q^0, u^0) \in V_i \subset H_1 \times H_2$ *ist genau dann Lösung der Aufgabe* (1), *wenn ein Element* $(l_1, l_2) \in H_1 \times H_2$ *existiert mit den Eigenschaften:*

$$\text{a)}\quad \|l_1\|_1 \leqq 1, \quad \|l_2\|_2 \leqq 1, \tag{3}$$

$$\text{b)}\quad (l_1, Q^0 - R)_1 + K(l_2, u^0)_2 = \|Q^0 - R\|_1 + K\,\|u^0\|_2, \tag{4}$$

$$\text{c)}\quad (l_1, Q^0 - Q)_1 + K\,(l_2, u^0 - u)_2 \leqq 0 \quad \forall (Q, u) \in V_i. \tag{5}$$

Beweis: Wurde vom ersten Autor unter Verwendung von Aussagen der Approximationstheorie in [11] geführt.

Satz 2: Die eindeutige Lösung der Aufgabe (2) bez. $V_1(Q^0, u^0)$ ergibt sich als Lösung der folgenden drei Bedingungen ($K > 0$):

$$(B^{\mathsf T} N^0, u^0)_2 - K\,\|u^0\|_2^2 = \operatorname*{Max}_{u \in U} \left((B^{\mathsf T} N^0, u)_2 - K\,\|u\|_2^2\right), \tag{6}$$

$$\frac{1}{2} A^{\mathsf T} N^0 = Q^0 - R, \qquad AQ^0 + Bu^0 + f = 0, \tag{7), (8}$$

wenn wir zusätzlich voraussetzen, daß N^0 zum Definitionsgebiet der zu A und B adjungierten Operatoren $A^{\mathsf T}$ und $B^{\mathsf T}$ gehört und (7) genau eine Lösung besitzt.

Beweis: Unter Verwendung der sogenannten Hamiltonfunktion (siehe [12])

$$H(N, Q, u) = (N, AQ + Bu + f)_3 - \|Q - R\|_1^2 - K\,\|u\|_2^2, \quad \frac{1}{2} A^{\mathsf T} N = Q - R \tag{9}$$

erhält man auf Grund der Konvexität des Funktionals aus (2)

$$J(u) - J(u^0) \geqq H(N^0, Q^0, u^0) - H(N^0, Q^0, u)$$

(siehe [12]). Aus (6) folgt $H(N^0, Q^0, u^0) - H(N^0, Q^0, u) \geqq 0$, d. h., u^0 ist Lösung der Aufgabe (2). Auf Grund der bereits oben festgestellten Einzigkeit der optimalen Steuerung, der eben gezeigten Hinlänglichkeit der Maximumbedingung (6) und der für ein $u^0 \in U$ existierenden eindeutigen Lösungen $N^0 = N(u^0)$, $Q^0 = Q(u^0)$ der Gleichungen (7) bzw. (8) ergibt sich nun die Behauptung des Satzes. ∎

Bemerkungen: 1. Satz 1 stellt eine Anwendung des KOLMOGOROW-Kriteriums der Approximationstheorie auf unsere Aufgabe (1) dar: Bei der Verwendung der Norm

$$\|(Q, u)\|_{1,2} := \|Q\|_1 + K\,\|u\|_2 \tag{10}$$

im Produktraum $H_1 \times H_2$ ist bei der Aufgabe (1) (für $K > 0$) das Element $(R, 0) \in H_1 \times H_2$ durch Elemente $(Q, u) \in V_i$ bez. dieser Norm (10) zu approximieren, d. h., unser Problem der optimalen Steuerung kann auch als Approximationsproblem gedeutet werden.

Satz 1 gilt auch noch allgemeiner in BANACHräumen (siehe [11]). Der Vorteil der Optimalitätsbedingungen aus diesem Satz liegt darin, daß sie im Unterschied zu Maximumprinzipien keinerlei Differenzierbarkeitsforderungen an das Funktional stellen. So ist ja bekannterweise die Norm $\|v\|$ für $v = 0$ selbst im HILBERTraum im Sinne von GÂTEAUX nicht differenzierbar, d. h., in diesen Fällen sind Maximumprinzipien nicht anwendbar.

2. Satz 2 stellt die Anwendung eines von einer Reihe von Autoren aufgestellten Maximumprinzips dar (siehe [12]). Für die Vergleichsmenge V_2^1 vereinfachen sich die Bedingungen (6)–(8) zu: u^0 ergibt sich als Lösung der Gleichung

$$L^{\mathsf{T}}(Lu^0 + r) + Ku^0 = 0, \quad \text{falls hieraus} \quad \|u^0\|_2 \leqq 1$$

folgt, ansonsten aus $u^0 = -L^{\mathsf{T}}(Lu^0 + r)/\|L^{\mathsf{T}}(Lu^0 + r)\|_2$.

3. Die Optimalitätsbedingungen (6)–(8) für die Aufgabe (2) eignen sich auf Grund ihrer komplizierten Struktur nur bei Spezialfällen (z. B. $U = H_2$) für eine numerische Bestimmung der optimalen Steuerung. Inwieweit sich die KOLMOGOROWbedingungen aus Satz 1 für eine numerische Lösung der Aufgabe (1) heranziehen lassen, bleibt noch zu untersuchen.

Kommen wir nun zu weiteren Eigenschaften der Aufgaben (1) und (2). Das folgende Lemma befaßt sich dabei mit der Abhängigkeit der Lösung der Aufgaben (1) und (2) von der Konstanten K.

Lemma 1: *Sei $u^0(K)$ eine Lösung der Aufgabe* (1) *bzw.* (2) *für die Vergleichsmenge V_1 (für ein festes $K \geqq 0$). Dann gilt ($i = 1, 2$) für $K_1 \leqq K_2$*

a) $J_i(u^0(K_1)) \leqq J_i(u^0(K_2))$,

b) $\|u^0(K_2)\|_2 \leqq \|u^0(K_1)\|_2$,

c) $\|Q(u^0(K_1)) - R\|_1 \leqq \|Q(u^0(K_2)) - R\|_1$

und für $K \to \infty$ $\|u^0(K)\|_2 \to 0$, falls $0 \in U$.

Beweis: Wird analog geführt wie der Beweis der entsprechenden Aussagen bei Straffunktionenmethoden, so daß wir hierauf verzichten.

Lemma 2: *Sei $u^0(K)$ Lösung der Aufgabe* (1) *bzw.* (2) *bez. V_1^1 oder V_2^1 mit $\|u^0(K)\|_2 = 1$ ($K > 0$ beliebig). Dann löst $u^0(K)$ auch die Aufgabe*

$$\|Q - R\|_1 \overset{!}{=} \underset{(Q,u) \in V_1^1 \text{ od. } V_2^1}{\text{Minimum}}.$$

Beweis: Für (1) bez. V_2^1 gilt $\|Lu^0(K)+r\|_1+K\,\|u^0(K)\|_2 \leqq \|Lu+r\|_1+K\,\|u\|_2$ $\forall u$ mit $\|u\|_2 \leqq 1$.
Hieraus folgt $\|Lu^0(K)+r\|_1 \leqq \|Lu+r\|_1+K\,(\|u\|_2-1) \leqq \|Lu+r\|_1$ $\forall u$ mit $\|u\|_2 \leqq 1$.
Für die anderen Aufgabenstellungen wird der Beweis analog geführt. ▮

Definition: Die Menge U sei äquilibriert. Eine Steuerung $u \in U$ heiße dann *Randsteuerung* von U, falls $\lambda u \notin U$ $\forall \lambda > 1$.

Man sieht leicht, daß die in Lemma 2 gegebene optimale Steuerung eine Randsteuerung im Sinne unserer Definition darstellt. Das folgende Lemma, das eine Verallgemeinerung von Aussagen aus [14] darstellt, gibt nun eine hinreichende Bedingung dafür an, wann eine optimale Steuerung Randsteuerung ist.

Lemma 3: *Sei u^0 optimal für die Aufgabe* (2) *bez.* V_2 *(für ein beliebiges, aber festes $K \geqq 0$). Wenn*

$$(\|L\|^2+K)\,M < |m|/2 \quad \text{und} \quad (Lu, r)_1 \neq 0 \quad \text{für ein } u \in U \quad \text{mit}$$

$$M = \sup_{u \in U} \|u\|_2^2 \quad \text{und} \quad m = \inf_{u \in U} (Lu, 2r)_1 = (Lv, 2r)_1$$

sind, dann ist u^0 Randsteuerung von U.

Beweis: Annahme, u^0 ist nicht Randsteuerung. Dann existiert ein $\varepsilon_0 > 0$, so daß $\lambda u^0 \in U$ $\forall \lambda$ mit $1 < \lambda < 1+\varepsilon_0$. Nun gilt für

$$\bar{J}(u) := \|Lu\|_1^2 + (Lu, 2r)_1 + K\,\|u\|_2^2:$$

$$\bar{J}(\lambda u^0) - \bar{J}(u^0) = (\lambda-1)\,\big((Lu^0, 2r)_1 + \lambda\,(\|Lu^0\|_1^2 + K\,\|u^0\|_2^2) + \|Lu^0\|_1^2 + K\,\|u^0\|_2^2\big)$$

und

$$\bar{J}(u^0) = (Lu^0, 2r)_1 + \|Lu^0\|_1^2 + K\,\|u^0\|_2^2 \leqq \bar{J}(v) \leqq m + (\|L\|^2+K)\,M$$
$$< m + |m|/(2+\varepsilon) = (1+\varepsilon)\,m/(2+\varepsilon) < 0 \quad \forall\, 0 < \varepsilon < \varepsilon_1$$

(da nach Voraussetzung $m < 0$ und $(\|L\|^2+K)\,M < |m|/(2+\varepsilon)$ $\forall\, 0 \leqq \varepsilon < \varepsilon_1$). Sei nun $2\varepsilon_0 < \varepsilon_1$. Wegen $\lambda\,(\|Lu^0\|_1^2 + K\,\|u^0\|_2^2) < (1+\varepsilon_0)\,|m|/(2+\varepsilon) = -m/2$ für $\varepsilon = 2\varepsilon_0$ folgt schließlich

$$\bar{J}(\lambda u^0) - \bar{J}(u^0) < \varepsilon_0\,\big((1+2\varepsilon_0)\,m/(2+2\varepsilon_0) - m/2\big) < 0 \quad \forall \lambda \text{ mit } 1 < \lambda < 1+\varepsilon_0,$$

wobei $2\varepsilon_0 < \varepsilon_1$. Dies steht im Widerspruch dazu, daß u^0 auch dem Funktional $\bar{J}(u)$ ein Minimum erteilt. ▮

Lemma 4: *Sei u^0 optimal für die Aufgabe* (2) *bez.* V_2^1 *(für ein beliebiges, aber festes $K \geqq 0$). Falls gilt*

$$K > 2\,\|L^\mathsf{T} r\|_2 > 0\,,$$

so ist u^0 keine Randsteuerung.

Beweis: Analog wie in Lemma 3 verwenden wir wieder das Funktional $\bar{J}(u) = \|Lu\|_1^2 + (Lu, 2r)_1 + K\,\|u\|_2^2$, das die gleichen Optimalpunkte wie $J(u)$ besitzt. Nun nehmen wir an, daß u^0 Randsteuerung ist, d. h. $\|u^0\|_2 = 1$, und bilden wieder den Ausdruck

$$\bar{J}(\lambda u^0) - \bar{J}(u^0) = (\lambda-1)\,\big((Lu^0, 2r)_1 + (\lambda+1)\,(\|Lu^0\|_1^2 + K\,\|u^0\|_2^2)\big)\,.$$

Wegen $(Lu^0, 2r)_1 + (\lambda+1)(\|Lu^0\|_1^2 + K\|u^0\|_2^2) \geqq -|(Lu^0, 2r)_1| + (\lambda+1)K \geqq -2\|L^{\mathsf{T}}r\|_2 + (\lambda+1)K > 2\lambda\|L^{\mathsf{T}}r\|_2 > 0$ erhalten wir für ein beliebiges $\lambda \in (0, 1)$ $\bar{J}(\lambda u^0) - \bar{J}(u^0) < 0$ im Widerspruch zur Optimalität von u^0. ▌

Bemerkungen: 1. Für die Vergleichsmenge V_2^1 vereinfacht sich die Bedingung $(\|L\|^2 + K)\, M < |m|/2$ aus Lemma 3 zu $(\|L\|^2 + K) < \|L^{\mathsf{T}}r\|_2$,

2. Unter zusätzlichen Voraussetzungen an den Operator L kann die Voraussetzung aus Lemma 4 abgeschwächt werden zu $K + E > 2\|L^{\mathsf{T}}r\|_2 > 0$ mit $0 < E = \operatorname{Inf}\{\lambda \mid \lambda \text{ Eigenwert von } L^{\mathsf{T}}L\}$, wie unmittelbar aus dem Beweis des Lemmas folgt.

3. Dualitätsaussagen

Kommen wir nun zu Dualitätsbeziehungen für unsere Aufgaben (1) und (2), wobei wir bekannte Dualitätstheorien verwenden. Das Ziel der folgenden Untersuchungen besteht nun darin, unter Verwendung dualer Aufgaben Schranken für unsere Funktionale zu berechnen und Lösungsmethoden zu entwickeln. Bei einigen der im weiteren betrachteten Aufgaben hat das duale Problem eine einfachere Struktur als das primale, so daß seine numerische Lösung effektiver ist als die numerische Behandlung des primalen Problems.

3.1. Eine Dualitätsaussage für die Aufgabe (1)

Satz 3: *Für die Aufgabe* (1) *bez.* V_1 *gilt die folgende Dualitätsbeziehung*

$$\underset{(Q,u)\in V_1}{\text{Minimum}} (\|Q-R\|_1 + K\|u\|_2) = \underset{\|l_1\|_1 \leqq 1, \|l_2\|_2 \leqq 1}{\text{Maximum}} ((l_1, Q^l - R)_1 + K(l_2, u^l)_2) \quad (11)$$

mit $(B^{\mathsf{T}}N - Kl_2, u^l)_2 = \underset{u\in U}{\text{Maximum}} (B^{\mathsf{T}}N - Kl_2, u)_2$ *und*

$$A^{\mathsf{T}}N = l_1, \quad AQ^l + Bu^l + f = 0 \quad (12)$$

(*unter den gleichen Voraussetzungen an die Operatoren wie in Satz* 2).

Beweis: Auf Grund der Beziehung $\|v\|_i = \underset{\|l\|_i \leqq 1}{\text{Maximum}} (l, v)_i$ und eines bekannten Minimaxsatzes (siehe [1]) erhalten wir

$$\underset{(Q,u)\in V_1}{\text{Minimum}} (\|Q-R\|_1 + K\|u\|_2) = \underset{(Q,u)\in V_1}{\text{Minimum}}\ \underset{\|l_1\|_1 \leqq 1, \|l_2\|_2 \leqq 1}{\text{Maximum}} ((l_1, Q-R)_1$$

$$+ K(l_2, u)_2) = \underset{\|l_1\|_1 \leqq 1, \|l_2\|_2 \leqq 1}{\text{Maximum}}\ \underset{(Q,u)\in V_1}{\text{Minimum}} ((l_1, Q-R)_1 + K(l_2, u)_2).$$

Die Minimierung des linearen Funktionals $(l_1, Q-R)_1 + K(l_2, u)_2$ bez. V_1 wird nun mit Hilfe des in [12] angegebenen Maximumprinzips durchgeführt. Dies ergibt die Beziehungen (12). ▌

Die Beziehung (11) zeigt, daß für die Vergleichsmenge V_1 die duale Aufgabe eine ziemlich komplizierte Struktur besitzt. Bei Verwendung der Vergleichsmenge V_2^1 vereinfacht sich die Dualitätsbeziehung (11) zu

$$\underset{\|u\|_2 \leqq 1}{\text{Minimum}} \left(\|Lu+r\|_1 + K\,\|u\|_2\right) = \underset{\|l_1\|_1 \leqq 1,\, \|l_2\|_2 \leqq 1}{\text{Maximum}} \left((l_1, r)_1 - \|L^{\mathrm{T}} l_1 + K l_2\|_2\right). \quad (11')$$

Unter Verwendung von (11′) wurden nun vom ersten Autor in [11] die folgenden Aussagen für die Aufgabe (1) bez. V_2^1 bewiesen: Für die optimale Steuerung u^0 erhält man im Falle $K > 0$ und $r \neq 0$ die Beziehung $u^0 = 0$, wenn die Ungleichung $\|L^{\mathrm{T}} r\|_2 / \|r\|_1 \leqq K$ erfüllt ist, während ansonsten (d. h. für $\|L^{\mathrm{T}} r\|_2/\|r\|_1 > K$) die Abschätzung $\|r\|_1 \geqq J_1(u^0) \geqq K\,\|r\|_1^2/\|L^T r\|_2 \geqq K\,\|r\|_1/\|L\| > 0$ gilt.

Für eine numerische Lösung wird man auch bei der Vergleichsmenge V_2^1 i. allg. nicht das duale Funktional verwenden, da es keine einfachere Struktur als das primale besitzt.

3.2. Dualitätsaussagen für die Aufgabe (2)

Im folgenden wenden wir die FENCHELsche und LAGRANGEsche Dualitätstheorie an, um Dualitätsbeziehungen für Probleme der Gestalt (2) zu erhalten. Dazu betrachten wir zuerst Aufgaben der Form

$$J(u) = \|u\|_2^2 \overset{!}{=} \underset{Lu=f}{\text{Minimum}} \quad \text{mit} \quad L \in L\,(H_2 \to H_3) \text{ und beschränkt}, \qquad (13)$$

auf die praktische Probleme der optimalen Steuerung mit festem Endpunkt führen, so z. B. die Aufgaben

$$\int_a^b u^2(t)\,\mathrm{d}t \overset{!}{=} \underset{u}{\text{Min.}} \quad \text{mit} \quad \dot{x} = A(t)\,x + B(t)\,u + f(t); \quad x(a) = x^a,\ x(b) = x^b$$

und

$$\int_0^a \int_0^b u^2(x, y)\,\mathrm{d}x\mathrm{d}y \overset{!}{=} \underset{u}{\text{Min.}} \quad \text{mit} \quad Q_{xy} = a(x, y)\,Q + b(x, y)\,Q_x + c(x, y)\,Q_y$$
$$+ d(x, y)\,u + e(x, y);$$
$$Q(x, 0) = f(x), \quad Q(0, y) = g(y), \quad Q(x, b) = F(x) \quad (\text{oder } Q(a, y) = G(y)),$$

für deren Zustandsgleichungen Lösungsformeln der Gestalt $x(t) = L(t)\,u + x_0$ bzw. $Q(x, y) = L(x, y)\,u + Q_0$ existieren. Setzt man hier nun die entsprechenden Endbedingungen ein, so ergibt sich eine Nebenbedingung der obigen Form $Lu = f$. Für das Problem (13) erhalten wir den folgenden Dualitätssatz

Satz 4: *Es gilt*

$$\underset{Lu=f}{\text{Minimum}}\, J(u) = \underset{Lu=f}{\text{Minimum}}\, \|u\|_2^2 = \underset{l \in H_3}{\text{Maximum}} \left(-\|L^{\mathrm{T}} l\|_2^2/4 + (l, f)_3\right) \qquad (14)$$
$$= \underset{l \in H_3}{\text{Maximum}}\, J_D(l).$$

Beweis: Unter Verwendung der LAGRANGEdualität erhalten wir

$$\underset{Lu=f}{\text{Minimum}}\, J(u) = \underset{l \in H_3}{\text{Maximum}}\, \underset{u \in H_2}{\text{Minimum}} \left(\|u\|_2^2 + (l, f - Lu)_3\right) \quad \text{(siehe [23])}.$$

Mit $\underset{u \in H_2}{\text{Minimum}} \left(\|u\|_2^2 + (l, f - Lu)_3\right) = -\|L^{\mathrm{T}} l\|_2^2/4 + (l, f)_3$ ergibt sich (14). ∎

Bemerkungen: 1. Satz 4 stellt eine Verallgemeinerung einer entsprechenden Aussage von Ulm [25] aus dem R^n dar.

2. Die duale Aufgabe aus Satz 4 ist für eine numerische Lösung besser geeignet als die primale Aufgabe (13), da wir es hier mit einem Problem ohne Beschränkungen zu tun haben. So führt z. B. schon das klassische Gradientenverfahren

$$l^{n+1}=l^n+\beta_n J'_D(l^n) \quad \text{mit} \quad J'_D(l)=-LL^{\mathsf{T}}l/2+f \tag{15}$$

bei entsprechender Schrittweitenwahl β_n zum Ziel.

3. Ist l^0 eine Lösung des dualen Problems, d. h. $-LL^{\mathsf{T}}l^0/2+f=0$, so ergibt sich $u^0=+L^{\mathsf{T}}l^0/2$ als Lösung des primalen Problems, wie man unmittelbar aus den Optimalitätsbedingungen für beide Probleme erhält.

Kommen wir nun zu allgemeinen Dualitätsaussagen für die Aufgabe (2').

Satz 5: *Im Falle $K>0$ gilt die folgende Dualitätsbeziehung für die Aufgabe* (2'):

$$\underset{u\in U}{\text{Minimum}}\,(\|Lu+r\|_1^2+K\,\|u\|_2^2)=\underset{y^*\in H_1}{\text{Maximum}}\,\big((P_U(L^{\mathsf{T}}y^*/2K), \tag{16}$$

$$KP_U(L^Ty^*/2K)-L^Ty^*)_2-(y^*,r+y^*/4)_1\big)\,.$$

Die optimale Lösung des dualen Problems berechnet sich dabei aus

$$LP_U(L^{\mathsf{T}}y^{0*}/2K)+\frac{1}{2}\,y^{0*}+r=0 \tag{17}$$

und die des primalen aus

$$u^0=P_U(L^{\mathsf{T}}y^{0*}/2K)\,, \tag{18}$$

wobei P_U den Projektionsoperator von H_2 auf U darstellt.

Beweis: Analog wie Mackenroth [8] bilden wir die Bifunktion

$$(Fy)\,(u):=\|Lu+y+r\|_1^2+K\,\|u\|_2^2\,.$$

Für die hierzu adjungierte Bifunktion ergibt sich

$$(F^*u^*)\,(y^*):=\underset{u\in U,y\in H_1}{\text{Infimum}}\,\big(\|Lu+y+r\|_1^2+K\,\|u\|_2^2-(u,u^*)_2+(y,y^*)_1\big)$$

$$=-\|y^*/2\|_1^2-(r,y^*)_1+\underset{u\in U}{\text{Infimum}}\,\big(K\,\|u\|_2^2-(u,u^*)_2-(u,L^{\mathsf{T}}y^*)_2\big)\,.$$

Hieraus folgt $(F^*0)\,(y^*)=-\|y^*\|_1^2/4-(r,y^*)_1-\|L^{\mathsf{T}}y^*\|_2^2/4K+K\,\|P_U(L^{\mathsf{T}}y^*/2K)-$ $-L^{\mathsf{T}}y^*/2K\|_2^2$. Auf Grund des für unsere Aufgabe gültigen Dualitätssatzes

$$\underset{u\in U}{\text{Minimum}}\,(F0)\,(u)=\underset{y^*\in H_1}{\text{Maximum}}\,(F^*0)\,(y^*) \qquad \text{(siehe [8])}$$

folgt die Beziehung (16). Infolge der für $f(v)=\underset{u\in U}{\text{Minimum}}\,\|u-v\|_2^2$ existierenden Fréchetableitung $f'(v)=2\,(I-P_U)\,v$ (siehe [16]), erhält man aus (16) unmittelbar die Optimalitätsbedingung (17). (18) folgt dann aus Charakterisierungssätzen (siehe [8]) für unsere Dualitätsbeziehung. ∎

Bemerkungen: 1. Dualitätsaussagen für die Aufgabe (2') wurden bisher nur von Mackenroth [8] und Mossino [6] aufgestellt. In Satz 5 haben wir die Vorgehensweise von MACKENROTH angewendet, wobei wir darüber hinaus durch Verwendung des Projektionsoperators eine duale Aufgabe „ohne Beschränkungen" erhielten, so daß wir die Gleichungen (17) und (18) zur Bestimmung der optimalen Größen y^{0*} und u^0 angeben konnten.

2. Für das Maximum des dualen Funktionals ergibt sich

$$\|y^{0*}\|_1^2/4 + K\,\|P_U(L^\mathsf{T} y^{0*}/2K)\|_2^2\,, \tag{19}$$

wie man unter Verwendung von (17) berechnet.

3. Für $K=0$ ist auch auf die Aufgabe (2′) der Dualitätssatz 3 anwendbar, der für alle $K \geqq 0$ Gültigkeit besitzt.

4. Gilt für die Aufgabe (2′) $U = H_2$, d. h., liegen keine Beschränkungen für u vor, so vereinfachen sich die Beziehungen (16)–(18) aus Satz 5 zu

$$\underset{u\in H_2}{\text{Minimum}}\,(\|Lu + r\|_1^2 + K\,\|u\|_2^2) = \underset{y^*\in \mathring{H}_1}{\text{Maximum}}\,(-\|y^*\|_1^2/4 - (r, y^*)_1 - \|L^\mathsf{T} y^*\|_2^2/4K)\,, \tag{16'}$$

$$LL^\mathsf{T} y^{0*}/2K + y^{0*}/2 + r = 0\,, \tag{17'}$$

$$u^0 = L^\mathsf{T} y^{0*}/2K\,. \tag{18'}$$

5. Liegen für u Beschränkungen der Form $\|u\|_2 \leqq 1$ vor, so berechnet sich der Projektionsoperator P_U bekanntlich aus

$$P_U(v) = \begin{cases} v\,, & \text{falls } \|v\|_2 \leqq 1 \\ v/\|v\|_2\,, & \text{falls } \|v\|_2 \geqq 1 \end{cases}\,,$$

so daß im Falle $\|L^\mathsf{T} y^{0*}/2K\|_2 \leqq 1$ die Beziehungen (17′) und (18′) gelten, während wir sonst

$$LL^\mathsf{T} y^{0*}/\|L^\mathsf{T} y^{0*}\|_2 + y^{0*}/2 + r = 0\,, \tag{17''}$$

$$u^0 = L^\mathsf{T} y^{0*}/\|L^\mathsf{T} y^{0*}\|_2 \tag{18''}$$

erhalten.

6. Satz 5 läßt sich auch analog für die Aufgabe (2) bez. V_1 formulieren, wenn die Operatoren A und B als beschränkt vorausgesetzt werden.

Im folgenden wenden wir nun den Dualitätssatz 5 an, um Schranken für das Funktional aus (2′) zu berechnen und ein Iterationsverfahren zur Lösung dieser Aufgabe abzuleiten. LEONDES und ARONOFF geben in [17] die Schranke

$$J_2^L(u) = \|Lu + r\|_1^2 + K\,\|u\|_2^2 \geqq K\,\|r\|_1^2/(\|L\|^2 + K) \quad \forall u \in H_2 \tag{20}$$

an, die sie unter Verwendung eines Projektionssatzes erhielten. Diese Schranke gilt natürlich auch, wenn Beschränkungen für u vorliegen. Unter Verwendung des Dualitätssatzes 5 läßt sich diese Schranke noch verbessern.

Lemma 5: *Für die Werte des Funktionals $J_2^L(u)$ aus (2′) gilt die folgende Ungleichung (unter der Voraussetzung $\|r\|_1 > 0$):*

$$J_2^L(u) \geqq K\,\|r\|_1^4/(K\,\|r\|_1^2 + \|L^\mathsf{T} r\|_2^2) \quad \forall u \in H_2\,. \tag{21}$$

Beweis: Setzen wir in dem dualen Funktional

$$-\|y^*\|_1^2/4-(r, y^*)_1-\|L^{\mathrm{T}}y^*\|_2^2/4K \quad \text{aus (16')} \quad y^*=-\beta r\ (\beta>0)\ ,$$

so folgt

$$F(\beta)=-\beta^2\,(\|r\|_1^2+\|L^{\mathrm{T}}r\|_2^2/K)/4+\beta\,\|r\|_1^2\,.$$

Die Maximierung von $F(\beta)$ bez. $\beta>0$ ergibt

$$\beta^0=2\,\|r\|_1^2/(\|r\|_1^2+\|L^{\mathrm{T}}r\|_2^2/K)\,.$$

Berechnet man $F(\beta^0)$ und wendet den Dualitätssatz 5 an, so folgt unmittelbar (21). ▌

Bemerkung: Die Schranke (21) ist wegen $\|L^{\mathrm{T}}r\|_2\leqq\|L\|\,\|r\|_1$ offensichtlich größer oder gleich der von Leondes und Aronoff gegebenen Schranke (20).

Unter zusätzlichen Voraussetzungen läßt sich bei der Beschränkung $\|u\|_2\leqq 1$ auch noch eine Schranke für das Funktional $J_2^L(u)$ ableiten, die besser als (20) ist.

Lemma 6: *Falls* $\|r\|_1\geqq\|L\|$ *ist und für die optimale Steuerung* u^0 $\|u^0\|_2=1$ *gilt, ergibt sich die folgende Schranke für das Funktional* $J_2^L(u)$ *bez. der Beschränkung* $\|u\|_2\leqq 1$:

$$J_2^L(u)\geqq K+(\|r\|_1-\|L\|)^2\ \forall u \quad \text{mit} \quad \|u\|_2\leqq 1\ , \tag{22}$$

wobei

$$K+(\|r\|_1-\|L\|)^2\geqq K\,\|r\|_1^2/(\|L\|^2+K) \tag{23}$$

gilt.

Beweis: Aus Satz 5 folgt unter Verwendung von (19)

$$J_2^L(u^0)=\|y^{0*}\|_1^2/4+K\,\|P_U(L^{\mathrm{T}}y^{0*}/2K)\|_2^2=\|y^{0*}\|_1^2/4+K\,.$$

Wegen $Lu^0+y^{0*}/2+r=0$ (siehe (17), (18)) erhalten wir

$$\|r\|_1=\|Lu^0+y^{0*}/2\|_1\leqq\|L\|+\|y^{0*}\|_1/2$$

und damit ergibt sich (22). (23) wird indirekt bewiesen: Die Annahme $K+(\|r\|_1-\|L\|)^2<K\cdot\|r\|_1^2/(\|L\|^2+K)$ führt zu $(K-\|L\|\,(\|r\|_1-\|L\|))^2<0$, d. h. zu einem Widerspruch. ▌

Lemma 7: *Die optimale Steuerung* u^0 *der Aufgabe* (2') *ist Lösung der folgenden Fixpunktgleichung:*

$$u^0=F(u^0):=P_U(-L^{\mathrm{T}}(Lu^0+r)/K)\ , \tag{24}$$

wobei der in (24) *definierte Operator* F *für* $\|L^{\mathrm{T}}L\|<K$ *kontraktiv und für* $\|L^TL\|=K$ *nichtexpansiv ist.*

Beweis: Die Fixpunktgleichung (24) ergibt sich unmittelbar aus (17) und (18), wenn man diese in der Form

$$y^{0*}/2=-Lu^0-r, \quad u^0=P_U(L^{\mathrm{T}}y^{0*}/2K)$$

schreibt.[4] Die Kontraktivität bzw. Nichtexpansivität von $F(u)$ folgt unmittelbar

[4] Dieselbe Fixpunktgleichung erhält man auch durch Anwendung der Optimalitätsbedingung $u^0=P_U\,(u^0-\lambda J'(u^0))$ für konvexe Funktionale.

aus

$$\|F(u^1)-F(u^2)\|_2=\|P_U(-L^T(Lu^1+r)/K)-P_U(-L^T(Lu^2+r)/K)\|_2$$
$$\leqq\|L^TL(u^1-u^2)\|_2/K\leqq\|L^TL\|\,\|u^1-u^2\|_2/K. \blacksquare$$

Bemerkung: Im Falle der Kontraktivität bzw. Nichtexpansivität des Operators F können nun Iterationsverfahren zur Bestimmung eines Fixpunktes von F und damit zur Bestimmung einer optimalen Steuerung angewendet werden.

Zusammenfassend kann gesagt werden, daß für die in dieser Arbeit betrachteten speziellen Probleme der optimalen Steuerung das duale Problem eine einfachere Struktur besitzt bzw. für eine numerische Behandlung Aussagen liefert, die aus dem primalen Problem nicht ersichtlich sind. In einer weiteren Arbeit wird die bisher in der Literatur wenig betrachtete numerische Auswertung von Dualitätssätzen für Steuerungsprobleme auf konkrete Aufgaben in Funktionenräumen angewandt.

Literatur

[1] Göpfert, A.: Mathematische Optimierung in allgemeinen Vektorräumen. Leipzig 1973.

[2] Mond, B., and M. Hanson: Duality for control problems. SIAM J. Control **6** (1968) 1, 114–120.

[3] Hager, W. W., and S. K. Mitter: Lagrange duality theory for convex control problems. J. Control and Optimization **14** (1976) 5, 843–856.

[4] Heins, W., and S. K. Mitter: Conjugate convex functions, duality and optimal control problems. Inf. Sciences **2** (1970), 211–243.

[5] Kuo, M. C. Y.: Duality in distributed parameter systems. J. of Eng. Math. **6** (1972) 3, 243–255.

[6] Mossino, J.: An application of duality to distributed optimal control problems with constraints on the control and the state. J. Math. Anal. Appl. **50** (1975), 223–242.

[7] Цветанов, М. М.: Двойственность в задачах вариационного исчисления и оптимального управления, Дисс. Москва 1970.

[8] Mackenroth, U.: Einige Bemerkungen zur Dualität in der konvexen Optimierung. Computing **14** (1975), 251–259.

[9] Mackenroth, U.: Dualität und Approximation bei parabolischen Kontrollproblemen. Diss. Mainz 1976.

[10] Ortlieb, C. P.: Dualität und Näherungsverfahren bei konvexen Steuerungsproblemen. Diss. Hamburg 1976.

[11] Benker, H.: Anwendung der Approximationstheorie auf ein spezielles Problem der optimalen Steuerung. Beitr. z. Analysis **14**.

[12] Benker, H., und E. Dittmar: Ein Maximumprinzip für Probleme der optimalen Steuerung I, II, III. Beitr. z. Analysis **7** (1975); **10** (1977); **12** (1978).

[13] Benker, H.: Anwendung von Methoden der nichtlinearen Optimierung auf Probleme der optimalen Steuerung in abstrakten Räumen. Beitr. z. Analysis **6** (1974), 37–47.

[14] Rogak, E. D., N. D. Kazarinoff and J. F. Scott-Thomas: Sufficient conditions for bang-bang control in Hilbert space, J. Opt. Theory Appl. **5** (1970) 1, 1–11.

[15] Иоффе, А. Д., и В. М. Тихомиров: Теория экстремальных задач. Москва 1974.
[16] Holmes, R. B.: A course on optimization and best approximation. Berlin, Heidelberg, New York 1972.
[17] Leondes, C. T., and E. Aronoff: Lower bounds for a quadratic cost functional. Int. J. Syst. Sci. **7** (1976), 17–25.
[18] Rockafellar, R. T.: Conjugate convex functions in optimal control and the calculus of variations. J. Math. Anal. Appl. **32** (1970), 174–222.
[19] Rockafellar, R. T.: Duality and stability in extremum problems involving convex functions. Pac. J. of Math. **21** (1967) 1, 167–187.
[20] Luenberger, D. G.: Optimization by vector space methods. New York 1969.
[21] Lions, J. L.: Optimal control of systems governed by partial differential equations. Berlin, Heidelberg, New York 1971.
[22] Glashoff, K., und W. Krabs: Dualität und Bang-Bang-Prinzip bei einem parabolischen Rand-Kontrollproblem, Bonner Math. Schriften **77** (1975), 1–8.
[23] Rolewicz, S.: Funktionalanalysis und Steuerungstheorie. Berlin, Heidelberg, New York 1976.
[24] Elster, K.-H., und R. Nehse: Zum Dualitätssatz von Fenchel. Math. Operationsforsch. Statist. **4** (1974) 5, 269–280.
[25] Ульм, С.: Об оптимизации на двух уровнях. Изв. ЭССР **1** (1969), 3–13.
[26] Krabs, W.: Zur Berechnung des Extremalwertes bei einem parabolischen Rand-Kontrollproblem. Beitr. z. Num. Mathem. **4** (1975), 129–139.
[27] Yavin, Y.: Optimal Control for a class of distributed parameter systems where the cost functions are norms. J. of Computer and System Sciences **3** (1969), 24–36.
[28] Yavin, Y., and R. Sivan: The bounded energy optimal control for a class of distributed parameter systems. Int. J. Control **8** (1968) 5, 525–536.
[29] Yavin, Y.: On the optimal control for a class of distributed parameter systems. Int. J. Control **12** (1970) 5, 889–895.

Summary

In this paper norm-minimal and quadratic control problems in Hilbert spaces are considered. For this problem optimality conditions and duality theorems are given. By application of these duality theorems the authors give lower bounds for the functionals and methods for solution of this problems.

Резюме

В статье рассматриваются минимальные по норме и квадратичные проблемы оптимального управления в гильбертовом пространстве. Для этих проблем доказываются условия оптимальности и теоремы двойственности. Эти теоремы двойственности применяются авторами для вычисления нижних границ функционалов и для решения этих проблем.

Eingereicht bei der Redaktion: 13. 1. 1978; revidierte Fassung: 14. 4. 1978

Math. Operationsforsch. Statist., Ser. Optimization, Vol. 10 (1979) No. 1, 143–162

On Optimal Control of MARKOV Processes with Application to a Class of Service Systems[1]

G. BIEBER[1]

Summary: A finite time control problem for MARKOV processes with arbitrary state and action space ist studied. The value of a given strategy is defined by the expected total reward up to a given point of time where the rewards are defined on the one hand by a reward rate and on the other hand by a valuation of the jumps of the controlled process. A necessary and sufficient condition for the existence of an optimal strategy is given. The structure of the studied processes enables us to interpret a class of optimization problems in queueing theory as special case of the general model.

1. Introduction

In the last years many authors have analysed problems of optimal control of service systems. The ask for the optimal planning of a service system is naturally inasmuch as the service system is a mathematical model for a real system.

Most results were obtained in the analysis of problems of optimal planning of priority queues. A summary of results for such problems can be found in SIMONOVITS [6].By the representation of results to the other problems of optimal control of service systems there are great difficulties. Above all this has its reason in the diversity of the models which were studied. To each of the manifold models of the queueing theory there are various possibilities to define optimization problems. A voluminous summary one can find in RYKOV [5].

Also FAINBERG, M. A., FAINBERG, E. A. [2], SOBEL [7], STIDHAM, PRABHU [8] give an impression of problems which were dealt with in the last years.

Most of models of control of service systems are models in which negativ exponentially distributed random variables (life times, interarrival times, ...) occur only. Such models are studied by the help of the theory of controllable MARKOV processes with discrete state space. More complicated models in which one random variable may be arbitrary distributed often are studied by the help of imbedded

[1] Hochschule für Architektur und Bauwesen Weimar, Sektion Rechentechnik und Datenverarbeitung, DDR – 53 Weimar, Karl-Marx-Platz.

semi-MARKOV processes. But it is clear that an optimal strategy for the imbedded model in general is not optimal for the original model.

In the present article we study a control problem for a general abstract class of service systems which was introduced in KÖNIG, MATTHES, NAWROTZKI [3], [4]. This extensive class of service systems which are called "service scheme" contains many of the known types of service systems. The problem which we study is a problem of optimal control of speeds for work.

For instance in special cases this can be a problem of optimal control of service rates.

The general conception of the speed for work is defined in section 5. The speed for work gives the speed of the realization of a random variable (for example: interarrival times, service times, ...). With the help of this conception one can introduce priorities too.

The control problem of section 5 is a problem of finite time control of a MARKOV process with general state space. In section 3 we define a control problem for MARKOV processes of the same structure as in section 5. In section 4 a necessary and sufficient condition for the existence of an optimal strategy is given.

This result is similar to that of VERMES [9]. The difference between our model and that of VERMES [9] above all consists in the reward structure.

In section 6 a simple example is given. In section 2 we state several notations and the concept of MARKOV processes which is used in the following sections.

I would like to thank Prof. D. KÖNIG for his helpful support and many valuable discussions.

2. Some fundamental notations

The concept of MARKOV processes used in this paper corresponds to DYNKIN [1]. Here we intend to state only some fundamental notations. Let be given some measurable spaces $[\Omega, \mathfrak{B}^0]$, $[E, \mathfrak{B}]$. A MARKOV process $\xi=(\xi_t, \zeta, \mathfrak{B}_t, P_x)$ on $[\Omega, \mathfrak{B}^0]$ with state space $[E, \mathfrak{B}]$ is given by

a) a function $\zeta\colon \Omega\to[0, \infty]$ called the life time of the process;
b) a function $\xi_t(\omega)=\xi(t, \omega)$ defined for each $\omega\in\Omega_t:=\{\omega : \zeta(\omega)>t\}$, $t\geqq 0$, taking values in the state space $[E, \mathfrak{B}]$;
c) a family $\{\mathfrak{B}_t\}_{t\geqq 0}$ of σ-algebras such that $\mathfrak{B}_t\subseteqq\mathfrak{B}^0$ is a σ-algebra on Ω_t for all $t\geqq 0$;
d) a family $\{P_x\}_{x\in E}$ of probability measures on $\mathfrak{B}^0$,

if the following conditions are satisfied:

e) $t\leqq u$ and $B\in\mathfrak{B}_t$ implies $B\cap\{\zeta>u\}\in\mathfrak{B}_u$;
f) $\{\xi_t\in C\}\in\mathfrak{B}_t$ for $t\geqq 0$, $C\in\mathfrak{B}$.
g) The function $P_x(\{\xi_t\in C\})$ is a $\mathfrak{B}$-measurable function of x for each $C\in\mathfrak{B}$ and all $t\geqq 0$;
h) $P_x(\{\xi_0\in E\setminus\{x\}\})=0$;

i) for arbitrary $t, s \geqq 0$, $C \in \mathfrak{B}$, $x \in E$
$P_x(\{\xi_{t+s} \in C\} \mid \mathfrak{B}_t) = P_{\xi_t}(\{\xi_s \in C\})$ a.s. Ω_t, P_x;

j) For each $t \geqq 0$, $\omega \in \Omega_t$, there exists an element $\omega_t \in \Omega_t$ such that

$$\xi_s(\omega_t) = \xi_{s+t}(\omega) \quad \text{and} \quad \zeta(\omega_t) = \zeta(\omega) - t$$

for all $0 \leqq s < \zeta(\omega) - t$.

By the help of ω_t it is possible to define a family $\{\Theta_t\}_{t \geqq 0}$ of operators

$$\Theta_t h(\omega) := h(\omega_t) \quad \text{for} \quad \omega \in \Omega_t$$

where h is an arbitrary function defined on E.

The set of all measurable bounded real functions defined on $[E, \mathfrak{B}]$ is denoted by $\boldsymbol{B}$.

The semigroup $\{\boldsymbol{T}_t\}_{t \geqq 0}$ of operators which are associated with a given Markov process is defined by

$$\boldsymbol{T}_t f(x) = \boldsymbol{E}_x f(\xi_t) \quad \text{for} \quad t \geqq 0, \quad f \in \boldsymbol{B}.$$

At this $\boldsymbol{E}_x$ denotes the expactation with respect to the measure $\boldsymbol{P}_x$.

A function f is said to be weak continuous with respect to $\{\boldsymbol{T}_t\}$ if

$$w\text{-}\lim_{t \searrow 0} \boldsymbol{T}_t f = f .$$

The weak infinitesimal operator $\boldsymbol{A}$ which is associated with a given Markov process is defined by

$$\boldsymbol{A} f = w\text{-}\lim_{t \searrow 0} \frac{\boldsymbol{T}_t f - f}{t}$$

for all functions $f \in \boldsymbol{B}$ for which this limit exists.

A function $f \in \boldsymbol{B}$ is called the weak limit of a sequence $\{f_n\}$ with $f_n \in \boldsymbol{B}$ for all n,

$$f = w\text{-}\lim_{n \to \infty} f_n$$

if for all $x \in E$ $f_n(x) \to f(x)$ as $n \to \infty$ and moreover $\|f_n\| = \sup_{x \in E} |f_n(x)|$ is a bounded sequence.

If ξ is a Markov process then a function $\tau \colon \Omega \to [0, \infty]$ is called a Markov time of the process ξ if

$$0 \leqq \tau(\omega) \leqq \zeta(\omega) \quad \text{for} \quad \omega \in \Omega$$

$$\{\tau > t\} \in \mathfrak{B}_t \qquad \text{for} \quad t \geqq 0 .$$

If τ is a Markov time of ξ then the measurable space $[\Omega_\tau, \mathfrak{B}_\tau]$ is defined by

$$\Omega_\tau := \{\omega \colon \tau(\omega) < \zeta(\omega)\}$$

and

$$B \in \mathfrak{B}_\tau \quad \text{iff} \quad B \subseteqq \Omega_\tau \quad \text{and} \quad B \cap \{\tau \leqq t < \zeta\} \in \mathfrak{B}_t$$

for all $t \geqq 0$.

A Markov process is said to be right-continuous if each function $\xi_t(\omega)$ for fixed $\omega \in \Omega_0$ is right-continuous for $t \in [0, \zeta(\omega))$.

We will use the following notations:

$$R^+ = \{x: x \in R^1, x \geqq 0\}$$

$\mathfrak{B}^+$: the σ-algebra of the BOREL-subsets of R^+

$\mathfrak{B}^{[s,t]}$: the σ-algebra of the BOREL-subsets of the interval $[s, t]$

$$[a]^+ = \begin{cases} a & \text{for } a \geqq 0 \\ 0 & \text{for } a < 0 \end{cases} \qquad \chi_M(a) = \begin{cases} 1 & \text{for } a \in M \\ 0 & \text{for } a \notin M . \end{cases}$$

3. Description of the model

Let be $\{\xi^\delta = (\xi_t, \zeta, \mathfrak{B}_t, P_x^\delta), \delta \in D\}$ a family of continuous on the right side homogenous MARKOV processes defined on a set Ω with state space $[E, \mathfrak{B}]$, whereby

$$E = R^+ \times A, \quad \mathfrak{B} = \mathfrak{B}^+ \otimes \mathfrak{A}$$

with a suitable set A.

$\mathfrak{A}$ is a σ-algebra generated by a topology on A which includes the one-point sets of A. About $[A, \mathfrak{A}]$ we assume that

$$A = \bigcup_{i=1}^{n} A_i$$

where the components A_i of A are separated by the topology on A. Because of this assumption the A_i are open as well as closed and belong to $\mathfrak{A}$.

The set D is a family of measurable functions (strategies) on $[[0, T) \times A, \mathfrak{B}^{[0,T)} \otimes \mathfrak{A}]$ with values in a measurable space $[V, \mathfrak{V}]$, called the decision-space, where the one-point sets of V belong to $\mathfrak{V}$. T is a positive finite real number, the duration of control.

For reasons of simplicity we define for the following

$$\begin{aligned} &E_i := R^+ \times A_i \\ &Z_i := [0, T) \times A_i \quad \text{for} \quad i = 1, \ldots, n \\ &Z := \bigcup_{i=1}^{n} Z_i . \end{aligned} \tag{3.1}$$

We make the following assumptions:

A1: For any $\delta_1, \delta_2 \in D$, $0 \leqq t < T$ the function

$$\delta(s, a) := \begin{cases} \delta_1(s, a) & 0 \leqq s < t, \quad a \in A \\ \delta_2(s, a) & t \leqq s < T, \quad a \in A \end{cases}$$

also belongs to D.

A2: If for $\delta_1, \delta_2 \in D$, $0 \leqq t_1 < t_2 < T$,

$$\delta_1(s, a) = \delta_2(s, a)$$

for each $a \in A$, $s \in [t_1, t_2)$, so the relation

$$P_{(s,a)}^{\delta_1}(\xi_t \in B) = P_{(s,a)}^{\delta_2}(\xi_t \in B)$$

is satisfied for all $(s, a) \in [t_1, t_2) \times A$, $B \in \mathfrak{B}$, $s + t \in [t_1, t_2]$.

A3: The process ξ^δ will be stopped at the time γ where

$$\gamma(\omega):=[\min\{\zeta(\omega), T-s\}]^+ \quad \text{for} \quad \xi_0(\omega)\in\{s\}\times A\,.$$

A4: The domains of the weak infinitesimal operators of the processes ξ^δ are independent of δ.

Regarding the special structure of E we interpret this model as mathematical model for any controllable system whose behaviour for fixed control strategy δ may be described by an (in general inhomogenous) MARKOV process with state space $[A, \mathfrak{A}]$. The additional time component in the description of the system is used only for simplification, because in that way it is possible to obtain a homogeneous MARKOV process.

Before we can introduce a reward structure and the control problem for our model we have to define some notations:

For $\omega\in\Omega_t=\{\omega:\zeta(\omega)>t\}$ and $0\leqq s\leqq t$ we define by

$$\xi_{[s,t]}(\omega):=\{\xi_h(\omega):s\leqq h\leqq t\}$$

the set of the values of the process in the time interval $[s, t]$ and for each $B\in\mathfrak{A}$, $\omega\in\Omega_0$ by

$$\tau_B(\omega):=\begin{cases}0 & \text{for} \quad \xi_0(\omega)\in R^+\times B\\ \sup\{t:\xi_{[0,t]}(\omega)\cap(R^+\times B)=\emptyset, \omega\in\Omega_t\} & \text{for} \quad \xi_0(\omega)\notin R^+\times B\end{cases}$$

the time until the process will enter the set $R^+\times B$. We call $g=g(\xi_t)$ the "coarse state" of the process at time t, $t\geqq 0$, where $g(x):=j$ for $x\in E_j$, $j=1,\ldots,n$. For $\omega\in\Omega_0$ we define

$$\tau_1(\omega):=\sup\Big\{t:\xi_{[0,t]}(\omega)\cap\Big(R^+\times\bigcup_{\substack{j=1\\ j\neq g_0(\omega)}}^{n}A_j\Big)=\emptyset,\ \omega\in\Omega_t\Big\}$$

and

$$\tau_i(\omega):=\begin{cases}\zeta(\omega) & \text{for} \quad \omega\in\Omega_{\tau_{i-1}(\omega)}\\ \sup\Big\{t:\xi_{[\tau_{i-1}(\omega),t]}(\omega)\cap\Big(R^+\times\bigcup\limits_{\substack{j=1\\ j\neq g_{\tau_{i-1}(\omega)}(\omega)}}^{n}A_j\Big)=\emptyset,\ \omega\in\Omega_t\Big\} & \text{for} \quad \omega\in\Omega_{\tau_{i-1}(\omega)}\end{cases}$$

for $i=2,3,\ldots$

τ_i is called the moment of the i-th change of the coarse state of the process, $i=1,2,\ldots$

Since the processes are continuous on the right side and since the A_j are open as well as closed and mutually disjoint the τ_i are MARKOV times, $i=1,2,\ldots$ (see DYNKIN [1], Lemma 4.1.). Let be $k^s_{(i,j)}(\omega)$ defined by

$$k^s_{(i,j)}(\omega):=\begin{cases}1 & \text{for} \quad (i,j)=(g_{\tau_{s-1}(\omega)}(\omega), g_{\tau_s(\omega)}(\omega)) \quad \text{and} \quad \omega\in\Omega_{\tau_s}\\ 0 & \text{otherwise}\end{cases}$$

for each $s=1,2,\ldots,(i,j)\in\Gamma$, where

$$\Gamma=\{(i,j):i,j=1,\ldots,n;\ i\neq j\},\quad \tau_0(\omega):\equiv 0\,.$$

(Clearly $\{k^s_{(i,j)}=1\}\in\mathfrak{B}_{\tau_s}$ for each $s=1,2,\ldots;\ (i,j)\in\Gamma$.)

We introduce by

$$s_t(\omega):=\sup\{i:\tau_i(\omega)<t\}\quad\text{for}\quad\omega\in\Omega_t$$

the number of changes of the coarse state until time t, $t\geqq 0$, and by

$$k_{(i,j)}(t,\omega):=\sum_{m=0}^{s_t(\omega)}k^m_{(i,j)}(\omega)\quad\text{for}\quad\omega\in\Omega_t$$

the number of changes of the coarse state from i to j until time t, $t\geqq 0$, and set formally $k^0_{(i,j)}(\omega):=0$, $\omega\in\Omega_0$, $(i,j)\in\Gamma$.

By usual transformations of these events it turns out that

$$\{k^{(t)}_{(i,j)}\geqq u\}\in\mathfrak{B}_t,\quad u=0,1,\ldots,+\infty\,.$$

Now we are in a position to introduce the following reward structure for our model.

Let be q a measurable bounded real function defined on $[Z\times V,\ \mathfrak{B}^{[0,T)}\otimes\mathfrak{A}\otimes\mathfrak{B}]$ and

$$q^\delta(x):=q(x,\delta(x))\quad\text{for}\quad x\in Z,\quad\delta\in D\,.$$

Furthermore let be h a measurable bounded real function defined on $[E\backslash Z,\ \mathfrak{B}^{[T,\infty)}\otimes\mathfrak{A}]$ and $\{k_{i,j}\}_{(i,j)\in\Gamma}$ a family of (finite) real numbers.

We define the value $\Phi(x,\delta)$ of a strategy $\delta\in D$ at a point $x\in E$ by

$$\Phi(x,\delta):=\boldsymbol{E}^\delta_x\left\{h(\xi_\gamma)+\sum_{(i,j)\in\Gamma}k_{(i,j)}(\gamma)\,k_{i,j}+\int_0^\gamma q^\delta(\xi_u)\,du\right\}$$

for $x\in Z$, and by

$$\Phi(x,\delta):=h(x)\quad\text{for}\quad x\in E\backslash Z\,,$$

where $\boldsymbol{E}^\delta_x$ means the expectation with respect to the measure $\boldsymbol{P}^\delta_x$ and the function γ is given in A3.

By this definition we see that the function h has the character of a valuation of the state at the moment where the process is stopped while q represents the cost rate for the valuation of the evolution of the process ξ^δ. $k_{i,j}$ gives the reward which arises always when a transition from the coarse state i to the coarse state j takes place, $(i,j)\in\Gamma$.

Definition 1: A strategy $\delta^*\in D$ is called *optimal* iff $\Phi(x,\delta^*)\leqq\Phi(x,\delta)$ for each $x\in Z$, $\delta\in D$.

4. A necessary and sufficient condition for optimality

For the further consideration we have to make the following additional assumptions:

A5: $\boldsymbol{E}^\delta_x k_{(i,j)}(t)=\begin{cases}o(t) & \text{for}\quad i\neq g(x)\\ \lambda^\delta_{xj}\cdot t+o(t) & \text{for}\quad i=g(x)\neq j\end{cases}$

for each $x\in Z$, $\delta\in D$, $(i,j)\in\Gamma$.

A6: The functions $q^\delta(x)$, $\lambda^\delta_{x,j}$ are weak continuous relative to the semigroup $\{T^\delta_t\}_{t\geqq 0}$ for each $\delta\in D$, $j=1, \dots, n$.

For simplification we set $\lambda^\delta_{x,g(x)}:=0$, for all $x\in Z$, $\delta\in D$.

Lemma: *For arbitrary $\delta\in D$, $x\in Z$*

$$A^\delta\Phi(x,\delta)+\sum_{j=1}^{n}\lambda^\delta_{x,j}k_{g(x),j}+q^\delta(x)=0$$

where A^δ denotes the weak infinitesimal operator of the process ξ^δ.

Proof: For fixed $\delta\in D$, $r\in\{1, \dots, n\}$, $x\in Z_r$ (for the definition of Z_r see (3.1)) we study the behaviour of

$$\frac{T^\delta_t\Phi(x,\delta)-\Phi(x,\delta)}{t}$$

for $t\searrow 0$.

We set $x=(s, a)$, $0\leqq s<T$, $a\in A_r$.

For all $t<T-s$ according to the definition of Φ we have

$$\begin{aligned} T^\delta_t\Phi(x,\delta)&=E^\delta_x\Phi(\xi_t,\delta)\\ &=E^\delta_x E^\delta_{\xi_t}\left\{h(\xi_\gamma)+\sum_{(i,j)\in\Gamma}k_{(i,j)}(\gamma)\,k_{i,j}+\int_0^\gamma q^\delta(\xi_u)\,du\right\},\end{aligned}$$

$$T^\delta_t\Phi(x,\delta)=E^\delta_x E^\delta_{\xi_t}h(\xi_\gamma)+E^\delta_x E^\delta_{\xi_t}\sum_{(i,j)\in\Gamma}k_{(i,j)}(\gamma)\,k_{i,j}+E^\delta_x E^\delta_{\xi_t}\int_0^\gamma q^\delta(\xi_u)\,du, \quad (3.2)$$

respectively.

Because of the definition of γ in A3 and theorem 3.1. in DYNKIN [1]

$$\begin{aligned} E^\delta_x E^\delta_{\xi_t}h(\xi_\gamma)&=E^\delta_x E^\delta_{\xi_t}h(\xi_{T-s-t})=E^\delta_x(\Theta_t h(\xi_{T-s-t}))\\ &=E^\delta_x h(\xi_{T-s-t}(\omega_t))=E^\delta_x h(\xi_{T-s})=E^\delta_x h(\xi_\gamma)\end{aligned} \quad (3.3)$$

$$\begin{aligned} E^\delta_x E^\delta_{\xi_t}\int_0^\gamma q^\delta(\xi_u)\,du&=\int_0^\infty E^\delta_x E^\delta_{\xi_t}(\chi_{[0,T-s-t]}(u)\,q^\delta(\xi_u))\,du \\ &=\int_t^\infty T^\delta_u(\chi_{[0,T-s-t]}(u-t)\cdot q^\delta)\,(x)\,du\\ &=E^\delta_x\int_t^{T-s}q^\delta(\xi_u)\,du\\ &=E^\delta_x\int_0^\gamma q^\delta(\xi_u)\,du-E^\delta_x\int_0^t q^\delta(\xi_u)\,du.\end{aligned} \quad (3.4)$$

By this transformations we have taken advantage of the boundedness of q^δ and of the theorem of FUBINI. Finally

$$E^\delta_x k_{(i,j)}(\gamma)=E^\delta_x k_{(i,j)}(t)+E^\delta_x E^\delta_{\xi_t}k_{(i,j)}\,(T-s-t)$$

and therefore

$$E^\delta_x E^\delta_{\xi_t}k_{(i,j)}(\gamma)=E^\delta_x k_{(i,j)}(\gamma)-E^\delta_x k_{(i,j)}(t)\,. \quad (4.4)$$

Putting (4.2), (4.3), (4.4) in (4.1) we obtain

$$\boldsymbol{T}_t^\delta \Phi(x, \delta) = \boldsymbol{E}_x^\delta h(\xi_\gamma) + \sum_{(i,j)\in\Gamma} \boldsymbol{E}_x^\delta k_{(i,j)}(\gamma)\, k_{i,j}$$

$$+ \boldsymbol{E}_x^\delta \int_0^\gamma q^\delta(\xi_u)\, du - \sum_{(i,j)\in\Gamma} \boldsymbol{E}_x^\delta k_{(i,j)}(t)\, k_{i,j}$$

$$- \boldsymbol{E}_x^\delta \int_0^t q^\delta(\xi_u)\, du$$

$$\boldsymbol{T}_t^\delta \Phi(x, \delta) = \Phi(x, \delta) - \sum_{(i,j)\in\Gamma} \boldsymbol{E}_x^\delta k_{(i,j)}(t)\, k_{i,j} - \boldsymbol{E}_x^\delta \int_0^t q^\delta(\xi_u)\, du\,. \tag{4.5}$$

By the definition of the weak infinitesimal operator

$$\boldsymbol{A}^\delta \Phi(x, \delta) = w\text{-}\lim_{t\searrow 0} \frac{\boldsymbol{T}_t^\delta \Phi(x, \delta) - \Phi(x, \delta)}{t}$$

and because of (4.5), A5, A6 we see that $\Phi(x, \delta)$ belongs to the domain of $\boldsymbol{A}^\delta$ and

$$\boldsymbol{A}^\delta \Phi(x, \delta) = w\text{-}\lim_{t\searrow 0} \frac{1}{t} \left(- \boldsymbol{E}_x^\delta \sum_{(i,j)\in\Gamma} k_{(i,j)}(t)\, k_{i,j} - \boldsymbol{E}_x^\delta \int_0^t q^\delta(\xi_u)\, du \right)$$

$$= - \sum_{j=1}^n \lambda_{x,j}^\delta k_{r,j} - q^\delta(x)$$

where r was defined above by $x \in Z_r$. The Lemma is proved. ∎

In the following theorem a sufficient and necessary condition for the existence of an optimal strategy is given.

Theorem 1: *There exists an optimal strategy if and only if there exists a measurable, bounded function $f\colon E \to R^1$ wich statisfies*

$$\min_{\delta\in D} \left(\boldsymbol{A}^\delta f(x) + \sum_{j=1}^n \lambda_{x,j}^\delta k_{r,j} + q^\delta(x) \right) = 0 \tag{4.6}$$

$$\textit{for} \quad x \in Z_r, \quad r = 1, \ldots, n, \quad \textit{and} \quad f(x) = h(x) \quad \textit{for} \quad x \in E \backslash Z\,. \tag{4.7}$$

If f is a solution of (4.6), (4.7) *and the minimum in* (4.6) *occurs for $\delta = \delta^*$, then δ^* is an optimal strategy.*

On the other hand if $\delta^* \in D$ is an optimal strategy and $f^*(x) := \Phi(x, \delta^*)$ then f^* is a solution of the problem (4.6), (4.7).

Proof: a) Let be $\delta^* \in D$ optimal, that is

$$\Phi(x, \delta^*) \leqq \Phi(x, \delta)$$

for arbitrary $x \in E$, $\delta \in D$. We show that the measurable bounded function

$$f^*(x) := \Phi(x, \delta^*) \quad \text{for} \quad x \in E$$

is a solution of the problem (4.6), (4.7). The proof will be indirect. We assume that f^* is not a solution of (4.6), (4.7), that is, there exist a strategy $\delta \in D$ and $x_0 = (s_0, a_0) \in \{s_0\} \times A_{r_0}$ for a certain number $r_0 \in \{1, \ldots, n\}$, $0 \leqq s_0 < T$, such that contrary to (4.6)

$$A^\delta f^*(x_0)+\sum_{j=1}^{n}\lambda^\delta_{x_0,j}k_{r_0,j}+q^\delta(x_0)<0\,. \tag{4.8}$$

Since $A^\delta f^*$, $\lambda^\delta_{x,j}$, $j=1,\ldots,n$, and q^δ are weak continuous relative to $\{T^\delta_t\}$ even there exists a $t_0>0$ such that

$$T^\delta_t\left(A^\delta f^*(x_0)+\sum_{j=1}^{n}\lambda^\delta_{x_0,j}k_{r_0,j}+q^\delta(x_0)\right)<0 \tag{4.9}$$

for all $0\leqq t<t_0$, where t_0 is to be less than $T-s_0$. From the inequality (4.9) it follows that

$$E^\delta_{x_0}\int_0^{t_0}\left(A^\delta f^*(\xi_u)+\sum_{j=1}^{n}\lambda^\delta_{\xi_u,j}k_{g(\xi_u),j}+q^\delta(\xi_u)\right)\mathrm{d}u<0\,. \tag{4.10}$$

Now define the strategy $\hat\delta$ by

$$\hat\delta(s,a):=\begin{cases}\delta^*(s,a) & \text{for}\quad s_0+t_0\leqq s<T,\quad a\in A\\ \delta(s,a) & \text{for}\quad 0\leqq s<s_0+t_0,\quad a\in A\end{cases} \tag{4.11}$$

and by assumption A1 we know that $\hat\delta\in D$. First of all we notice

$$\begin{aligned}\Phi(x_0,\delta^*)&=f^*(x_0)\\ &=T^{\hat\delta}_{t_0}f^*(x_0)-\int_0^{t_0}T^{\hat\delta}_uA^{\hat\delta}f^*(x_0)\,\mathrm{d}u\\ &=E^{\hat\delta}_{x_0}f^*(\xi_{t_0})-E^{\hat\delta}_{x_0}\int_0^{t_0}A^{\hat\delta}f^*(\xi_u)\,\mathrm{d}u\,.\end{aligned} \tag{4.12}$$

(4.12) follows from assertion 1.15.C. in DYNKIN [1]. Since $T-s_0>t_0$ $\Phi(x_0,\hat\delta)$ may be represented by

$$\begin{aligned}\Phi(x_0,\hat\delta)=E^{\hat\delta}_{x_0}\{h(\xi_\gamma)&+\sum_{(i,j)\in\Gamma}k_{(i,j)}(\gamma)\,k_{i,j}\\ &+\int_{t_0}^{\gamma}q^{\hat\delta}(\xi_u)\,\mathrm{d}u\}+E^{\hat\delta}_{x_0}\int_0^{t_0}q^{\hat\delta}(\xi_u)\,\mathrm{d}u\,.\end{aligned} \tag{4.13}$$

For the first part of the right side of (4.13) we find analogous as in the proof of the lemma

$$\begin{aligned}\Phi(x_0,\hat\delta)-E^{\hat\delta}_{x_0}\int_0^{t_0}q^{\hat\delta}(\xi_u)\,\mathrm{d}u&=E^{\hat\delta}_{x_0}E^{\hat\delta}_{\xi_{t_0}}h(\xi_\gamma)\\ &+E^{\hat\delta}_{x_0}E^{\hat\delta}_{\xi_{t_0}}\int_0^{\gamma}q^{\hat\delta}(\xi_u)\,\mathrm{d}u+E^{\hat\delta}_{x_0}\sum_{(i,j)\in\Gamma}k_{(i,j)}(\gamma)\,k_{i,j}\end{aligned} \tag{4.14}$$

$$\begin{aligned}&=E^{\hat\delta}_{x_0}E^{\delta^*}_{\xi_{t_0}}h(\xi_\gamma)+E^{\hat\delta}_{x_0}E^{\delta^*}_{\xi_{t_0}}\int_0^{\gamma}q^{\delta^*}(\xi_u)\,\mathrm{d}u\\ &\quad+E^{\hat\delta}_{x_0}\sum_{(i,j)\in\Gamma}k_{(i,j)}(\gamma)\,k_{i,j}=E^{\hat\delta}_{x_0}f^*(\xi_{t_0})\\ &\quad+E^{\hat\delta}_{x_0}\sum_{(i,j)\in\Gamma}k_{(i,j)}(\gamma)\,k_{i,j}-E^{\hat\delta}_{x_0}E^{\delta^*}_{\xi_{t_0}}\sum_{(i,j)\in\Gamma}k_{(i,j)}(\gamma)\,k_{i,j}\end{aligned}$$

where we have taken in consideration the definition (4.11) of $\hat{\delta}$ and assumption A2. Also from 1.15.C. in Dynkin [1] we find that

$$\boldsymbol{E}_{x_0}^{\hat{\delta}} \sum_{(i,j)\in\Gamma} k_{(i,j)}(\gamma)\, k_{i,j} = \sum_{(i,j)\in\Gamma} \boldsymbol{E}_{x_0}^{\hat{\delta}} \boldsymbol{E}_{\xi_{t_0}}^{\hat{\delta}} k_{(i,j)}(\gamma)\, k_{i,j} - \sum_{(i,j)\in\Gamma} \boldsymbol{E}_{x_0}^{\hat{\delta}} \int_0^{t_0} \boldsymbol{A}^{\hat{\delta}} \boldsymbol{E}_{\xi_u}^{\hat{\delta}} k_{(i,j)}(\gamma)\, k_{i,j}\, du \,. \tag{4.15}$$

Now analogous as in the proof of the lemma (see (4.4)) for the integrand in (4.15) we obtain

$$\boldsymbol{A}^{\hat{\delta}} \boldsymbol{E}_{\xi_u}^{\hat{\delta}} k_{(i,j)}(\gamma) = w\text{-}\lim_{t\searrow 0} \frac{1}{t} \boldsymbol{E}_{\xi_u}^{\hat{\delta}} [\boldsymbol{E}_{\xi_t}^{\hat{\delta}} k_{(i,j)}(\gamma) - k_{(i,j)}(\gamma)]$$

$$= w\text{-}\lim_{t\searrow 0} \left(-\frac{1}{t} \boldsymbol{E}_{\xi_u}^{\hat{\delta}} k_{(i,j)}(t) \right) = \begin{cases} 0 & \text{for } \; i \neq g(\xi_u) \\ -\lambda_{\xi_u,j}^{\hat{\delta}} & \text{for } \; i = g(\xi_u)\,. \end{cases}$$

Putting this in (4.15) we obtain

$$\boldsymbol{E}_{x_0}^{\hat{\delta}} \sum_{(i,j)\in\Gamma} k_{(i,j)}(\gamma)\, k_{i,j} = \sum_{(i,j)\in\Gamma} \boldsymbol{E}_{x_0}^{\hat{\delta}} \boldsymbol{E}_{\xi_{t_0}}^{\hat{\delta}} k_{(i,j)}(\gamma)\, k_{i,j} + \boldsymbol{E}_{x_0}^{\hat{\delta}} \sum_{j=1}^{n} \int_0^{t_0} \lambda_{\xi_u,j}^{\hat{\delta}} k_{g(\xi_u),j}\, du \tag{4.16}$$

and together with (4.14) as well as the definition (4.11) of $\hat{\delta}$ for the value of $\hat{\delta}$ we have the equation

$$\Phi(x_0, \hat{\delta}) = \boldsymbol{E}_{x_0}^{\hat{\delta}} f^*(\xi_{t_0}) + \boldsymbol{E}_{x_0}^{\hat{\delta}} \int_0^{t_0} q^{\hat{\delta}}(\xi_u)\, du + \boldsymbol{E}_{x_0}^{\hat{\delta}} \sum_{j=1}^{n} \int_0^{t_0} \lambda_{\xi_u,j}^{\hat{\delta}} k_{g(\xi_u),j}\, du$$

$$= \boldsymbol{E}_{x_0}^{\hat{\delta}} f^*(\xi_{t_0}) + \boldsymbol{E}_{x_0}^{\hat{\delta}} \int_0^{t_0} q^{\hat{\delta}}(\xi_u)\, du + \boldsymbol{E}_{x_0}^{\hat{\delta}} \int_0^{t_0} \sum_{j=1}^{n} \lambda_{\xi_u,j}^{\hat{\delta}} k_{g(\xi_u),j}\, du \,.$$

Considering else (4.12) from the last equation it follows that

$$\Phi(x_0, \hat{\delta}) - \Phi(x_0, \delta^*) = \boldsymbol{E}_{x_0}^{\hat{\delta}} \int_0^{t_0} \left[\boldsymbol{A}^{\delta} f^*(\xi_u) + \sum_{j=1}^{n} \lambda_{\xi_u,j}^{\delta} k_{g(\xi_u),j} + q^{\delta}(\xi_u) \right] du$$

and because of (4.9) finally

$$\Phi(x_0, \hat{\delta}) - \Phi(x_0, \delta^*) < 0$$

contradictory to the optimality of δ^*.

b) Let now be f^* a (measurable bounded) solution of (4.6), (4.7) and let be $\delta^* \in D$ a strategy for which the minimum in (4.6) occurs. We show that δ^* is optimal. First of all from 1.15.C in Dynkin [1] it follows that for $x \in \{s\} \times A$, $0 \leqq s < T$,

$$f^*(x) = \boldsymbol{E}_x^{\delta} f^*(\xi_t) - \boldsymbol{E}_x^{\delta} \int_0^{t} \boldsymbol{A}^{\delta} f^*(\xi_u)\, du \tag{4.17}$$

for each $\delta \in D$, $t \geqq 0$. In particular since f^* is a solution of (4.6), (4.7) we see that

$$f^*(x) = \boldsymbol{E}_x^{\delta^*} f^*(\xi_t) + \boldsymbol{E}_x^{\delta^*} \int_0^{t} \left(\sum_{j=1}^{n} \lambda_{\xi_u,j}^{\delta^*} k_{g(\xi_u),j} + q^{\delta^*}(\xi_u) \right) du \,. \tag{4.18}$$

As the part a) of the present proof (see (4.16)) shows moreover

$$\boldsymbol{E}_x^{\delta^*}\int_0^t\left(\sum_{j=1}^n \lambda_{\xi_u,j}^{\delta^*}k_{g(\xi_u),j}\right)\mathrm{d}u \tag{4.19}$$
$$=\boldsymbol{E}_x^{\delta^*}\sum_{(i,j)\in\Gamma}k_{(i,j)}(\gamma)\,k_{i,j}-\sum_{(i,j)\in\Gamma}\boldsymbol{E}_x^{\delta^*}\boldsymbol{E}_{\xi_t}^{\delta^*}k_{(i,j)}(\gamma)\,k_{i,j}\,.$$

The second part of the right side of (4.19) vanishes for $t=T-s$. Therefore (4.18) and (4.19) give

$$f^*(x)=\boldsymbol{E}_x^{\delta^*}f^*(\xi_{T-s})+\boldsymbol{E}_x^{\delta^*}\sum_{(i,j)\in\Gamma}k_{(i,j)}(\gamma)\,k_{i,j}+\boldsymbol{E}_x^{\delta^*}\int_0^{T-s}q^{\delta^*}(\xi_u)\,\mathrm{d}u \tag{4.20}$$
$$=\boldsymbol{E}_x^{\delta^*}h(\xi_\gamma)+\boldsymbol{E}_x^{\delta^*}\sum_{(i,j)\in\Gamma}k_{(i,j)}(\gamma)\,k_{i,j}+\boldsymbol{E}_x^{\delta^*}\int_0^{\gamma}q^{\delta^*}(\xi_u)\,\mathrm{d}u=\Phi(x,\delta^*)\,.$$

On the other hand for each strategy $\delta\in D$

$$f^*(x)=\boldsymbol{E}_x^{\delta}f^*(\xi_{T-s})-\boldsymbol{E}_x^{\delta}\int_0^{T-s}\boldsymbol{A}^{\delta}f^*(\xi_u)\,\mathrm{d}u \tag{4.21}$$

$$f^*(x)\leqq\boldsymbol{E}_x^{\delta}h(\xi_\gamma)+\boldsymbol{E}_x^{\delta}\int_0^{T-s}\left(\sum_{j=1}^n\lambda_{\xi_u,j}^{\delta}k_{g(\xi_u),j}+q^{\delta}(\xi_u)\right)\mathrm{d}u \tag{4.22}$$

$$f^*(x)\leqq\boldsymbol{E}_x^{\delta}h(\xi_\gamma)+\boldsymbol{E}_x^{\delta}\sum_{(i,j)\in\Gamma}k_{(i,j)}(\gamma)\,k_{i,j}+E_x^{\delta}\int_0^{\gamma}q^{\delta}(\xi_u)\,\mathrm{d}u\,, \tag{4.23}$$

that is

$$f^*(x)\leqq\Phi(x,\delta)\,. \tag{4.24}$$

The equation (4.20) holds by (4.17) and (4.22) follows because f^* is a solution of (4.6), (4.7) and the minimum in (4.6) is attained for δ^*.

The inequality (4.22) is obtained from (4.22) by analogous transformations as in the proof of (4.19).

Finally (4.20) and (4.24) give

$$\Phi(x,\delta^*)\leqq\Phi(x,\delta)$$

for each $\delta\in D$. The theorem is proved.

Now let the set of decisions V be given by

$$V=\bigcup_{i=1}^n V_i$$

where $V_i\cap V_j=\emptyset$ for $i\neq j$. We interpret the set V_i as the set of admissible decisions if the process is in a state $x\in Z_i$, $i=1,\ldots,n$. In the following we will need the conditions

(I) For an arbitrary family $(v_i)_{i=1,2,\ldots,n}$ of decisions such that $v_i\in V_i$, $i=1,\ldots,n$, the function δ defined by

$$\delta(x)=v_{g(x)}\quad\text{for}\quad x\in Z$$

belongs to D.

(II) If $\delta_1, \delta_2 \in D$ and for some $x_0 \in Z$

$$\delta_1(x_0) = \delta_2(x_0)$$

is valid then

$$A^{\delta_1} f(x_0) = A^{\delta_2} f(x_0) \qquad (4.25)$$
$$\lambda^{\delta_1}_{x_0,j} = \lambda^{\delta_2}_{x_0,j} \qquad \text{for} \quad j = 1, \ldots, n$$

also is valid for each f which belongs to the domain of A^{δ}, $\delta \in D$. For reasons of simplicity we choose a fixed family $(v_i)_{i=1,\ldots,n}$ of admissible decisions and define for every decision $v \in V_i$ the function (v) by

$$(v)(x) := v \qquad \text{for} \quad x \in Z_i$$
$$(v)(x) := v_{g(x)} \quad \text{for} \quad x \notin Z_i ,$$

$i = 1, \ldots, n$. If assumption (I) is valid then always $(v) \in D$.

Theorem 2: *If the conditions* (I), (II) *are satisfied then there is an optimal strategy if and only if there is a strategy $\delta \in D$ and a* (*measurable bounded*) *function $f: E \to R^1$ such that*

$$A^{(\delta(x))} f(x) + \sum_{j=1}^{n} \lambda^{(\delta(x))}_{x,j} k_{g(x),j} + q^{(\delta(x))}(x) = 0 \quad \text{for} \quad x \in Z \qquad (4.26)$$

$$A^{(v)} f(x) + \sum_{j=1}^{n} \lambda^{(v)}_{x,j} k_{g(x),j} + q^{(v)}(x) \geqq 0 \quad \text{for} \quad x \in Z, \quad v \in V_{g(x)} \qquad (4.27)$$

$$f(x) = h(x) \quad \text{for} \quad x \in E \backslash Z . \qquad (4.28)$$

If δ, f solve (4.24), (4.27), (4.28) *then δ is an optimal strategy and $f(x) = \Phi(x, \delta)$.*

Remarks: 1. If there exists some measurable bounded function f which satisfies (4.27) and for each $x \in Z$ there is a set $V(x)$, $\emptyset \neq V(x) \subseteqq V_{g(x)}$ such that the equality in (4.26) is realized for each $\delta(x) = (v)$, $v \in V(x)$ then (4.26) secures the existence of a measurable function $d: Z \to V$ such that

$$d(x) \in V(x) \quad \text{for each} \quad x \in Z .$$

This becomes evident by the following condition which is equivalent to (4.26), (4.27)

$$\min_{v \in V_{g(x)}} \left[A^{(v)} f(x) + \sum_{j=1}^{n} \lambda^{(v)}_{x,j} k_{q(x),j} + q^{(v)}(x) \right] \qquad (4.29)$$
$$= A^{(\delta(x))} f(x) + \sum_{j=1}^{n} \lambda^{(\delta(x))}_{x,j} k_{g(x),j} + q^{(\delta(x))}(x) = 0 .$$

2. In the proof of theorem 2 we can see that (4.25) of condition (II) must be valid only for a solution f^* of (4.26), (4.27), (4.28). Therefore if condition (I) is valid a given strategy $\delta^* \in D$ for which $f^* = \Phi(\cdot, \delta^*)$ satisfies (4.25) is optimal if and only if δ^*, f^* satisfy (4.26), (4.27).

Proof of theorem 2: a) Let be $\delta^* \in D$, f^* satisfy (4.26), (4.27), (4.28). Then for arbitrary $\delta \in D$ by (II)

$$A^{\delta}f^*(x)+\sum_{j=1}^{n} \lambda_{x,j}^{\delta} k_{g(x),j}+q^{\delta}(x)$$

$$=A^{(\delta(x))}f^*(x)+\sum_{j=1}^{n} \lambda_{x,j}^{(\delta(x))} k_{g(x),j}+q^{(\delta(x))}(x)\geqq 0$$

and analogous

$$A^{\delta^*}f^*(x)+\sum_{j=1}^{n} \lambda_{x,j}^{\delta^*} k_{g(x),j}+q^{\delta^*}(x)=0 .$$

Therefore f^* is a solution of (4.6), (4.7) and the minimum in (4.6) is obtained for $\delta=\delta^*$. Consequently, δ^* is an optimal strategy.

b) Let be $\delta^* \in D$ an optimal strategy and $f^*(x):=\Phi(x, \delta^*)$. Then because of the Lemma and by (II) it follows that δ^*, f^* satisfy (4.26). If now there exist $x_0 \in Z$, $v_0 \in V_{g(x_0)}$ such that

$$A^{(v_0)}f^*(x_0)+\sum_{j=1}^{n} \lambda_{x_0,j}^{(v_0)} k_{g(x_0),j}+q^{(v_0)}(x_0)<0$$

then since $(v_0) \in D$ by theorem 1 we get a contradiction to the optimality of δ^*. The theorem is proved. ▌

5. Control of the speeds for work for a class of service systems

Let be given a service scheme by the quantities J, $\{C_j, D_j\}_{j\in J}$, G, $\{p(g, s, g')\}_{g,g'\in G, s\in J_g}$. For a detailed representation of this conception see KÖNIG, MATTHES, NAWROTZKI [3], [4]. We interpret such a model as follows: A system consists of finitely many objects O_j, $j \in J$ (for example stream of customers). To each object O_j, $j\in J$, belong the sets C_j and D_j of states such that

$$C_j \cap D_j = 0, \quad C_j \cup D_j \neq 0 .$$

C_j is called the set of active states of O_j and D_j the set of passive states of O_j.

A coarse state of the system will be described by a family $g=(i_j)_{j\in J} \in G \subseteqq \bigtimes_{j\in J} (C_j \cup D_j)$. In this way G denotes the set of all coarse states of the system. We assume that $G \neq 0$.

Changes of the coarse state of the system take place iff any object finishes its active state. $p(g, s, g')$ gives the probability that the new coarse state of the system will be g' if g is the actual coarse state and the object O_j finishes its active state, $i_s \in C_s$, $g=(i_j)_{j\in J}\in G$, $g' \in G$:

$$\begin{aligned}
&p(g, s, g') \geqq 0 && \text{for} \quad g' \in G, \quad i_s \in C_s, \quad g=(i_j)_{j\in J} \in G\\
&\sum_{g'\in G} p(g, s, g')=1 && \text{for} \quad g=(i_j)_{j\in J}\in G, \quad i_s \in C_s\\
&p(g, s, g')=0 && \text{for} \quad g=(i_j)_{j\in J}\in G, \quad g'=(i_j')_{j\in J}\in G\\
& && \qquad i_j\in C_j, \quad i_j \neq i_j', \quad i_s \in C_s, \quad s\neq j .
\end{aligned}$$

Without loss of generality we assume that a direct transition from g to g is impossible for each $g \in G$:

$$p(g, s, g) = 0 \quad \text{for each} \quad g \in G, \quad s \in J_g, \tag{5.1}$$

where

$$J_g := \{j : i_j \in C_j\} \quad \text{for} \quad g = (i_j)_{j \in J} \in G .$$

If the condition (5.1) for a given model is not valid so one can always find an other model which satisfies (5.1) and is equivalent to the original model (see also the example). We associate a distribution density $f_{i_j,j}(x)$ of a positive random variable to each active state $i_j \in C_j$ of the object O_j and denote by $F_{i_j,j}(x)$ the corresponding distribution function. We interpret $F_{i_j,j}(x)$ as distribution function of the random quantity of work which the object O_j has to perform if O_j is in the active state i_j. O_j finishes its active state i_j just in the moment when this object has performed the complete quantity of work.

We assume that all this random variables are independent of each other.

The fine state of an object O_j which is in an active state $i_j \in C_j$ is defined by a pair

$$a_j := [i_j, y_j] ,$$

where the additional variable y_j denotes the quantity of work which the object has performed previously, $F_{i_j,j}(y_j) < 1$. If the state of O_j is passive, $i_j \in D_j$, so we define

$$a_j := i_j .$$

According to that we denote the (fine) state of the system by the family $a = (a_j)_{j \in J}$ of fine states of the objects and lump all possible (fine) states of the system for fixed coarse state $g \in G$ to the set A_g. Then $A = \bigcup_{g \in G} A_g$ is the set of all possible states of the system which we by transfer of the topology from R^+ expand to the measurable state space $[A, \mathfrak{A}]$. (The notation in this section is chosen in complete anology to that of the section 3 so far as possible.) We associate a set V_{jg} to each pair (j, g) for which $g \in G, j \in J_g$ such that

$$\emptyset \neq V_{jg} \subseteq R^+$$

and a measurable bounded function b_{jg}: $V_{jg} \to R^1$. V_{jg} denotes the set of admissible speeds for work of the object O_j if the system is in a state $a \in A_g$ and $b_{jg}(v)$ represents the cost rate at the object O_j when it works with speed $v \in V_{j,g}$.

Since in a real system a bound for the total capacity possibly exists we introduce sets V_g for which

$$\emptyset \neq V_g \subseteq \bigtimes_{j \in J_g} V_{jg} \quad \text{for} \quad g \in G$$

and define by $V = \bigcup_{g \in G} V_g$ the set of decisions. Analogous to the enlargement of the set of states A we expand the set V in natural way to a measurable (decision) space $[V, \mathfrak{B}]$.

Additional to the reward which depends on the speeds of work we admit the following rewards:

$k_{g,g'}$: the reward which obtained always when in the system a transition from a state of A_g to a state of A takes place, $g \neq g'$, $g \in G$, $g' \in G$;

(In a real system this reward possibly depends on the number of the object which produces this transition. But always we are able to transform such a model to a model in which the rewards are as above.)

$w_{i,j}$: the reward which is obtained in any time unit when the object O_j is in the passiv state $i \in D_j$, $j \in J$;

$h(a)$: h is a measurable bounded real function and $h(a)$ denotes the reward which is obtained when the system is stopped in the state a, $a \in A$.

We set

$$w_g = \sum_{j \in J \setminus J_g} w_{i_j,j} \qquad \text{for} \quad g = (i_j)_{j \in J} \in G$$

$$b_g(\mathfrak{v}) = \sum_{j \in J_g} b_{j,g}(v_{j,g}) \qquad \text{for} \quad \mathfrak{v} = (v_{j,g})_{g \in G, j \in J_g}$$

$$q(t, a, \mathfrak{v}) = b_g(\mathfrak{v}) + w_g \quad \text{for} \quad t \geqq 0, \quad g \in G, \quad a \in A_g$$

and notice that all the following statements remain true if w_g respectively $b_g(\mathfrak{v})$ are composed not additiv, but are any real numbers respectively measurable bounded functions. Likewise, it might be that all rewards are dependent on the actual time. But for simplicity we renounce such a representation. A function δ defined on $[[0, T) \times A, \mathfrak{B}^+ \otimes \mathfrak{A}]$ with values in $[V, \mathfrak{B}]$ is called an admissible strategy if δ is measurable and bounded in each of its components and if

$$\delta(t, a) \in V_g \quad \text{for} \quad a \in A_g, \quad g \in G, \quad t \in [0, T).$$

Let δ be an admissible strategy. We say that our system is controlled by the strategy δ if for each $g \in G$, $a \in A_g$, $j \in J_g$ the speed of work of the object O_j at time t is defined by the j-component of $\delta(t, a)$ (which we will denote by $\delta_j(t, a)$) if the state of the system at time t is $a \in A$. We notice that the (random) temporal progress of the state of our system is well defined if an admissible strategy for the system is given. By the complex description of the states and since each admissible strategy is independent on the past the temporal progress of the controlled system is described by an (in general inhomogeneous) MARKOV process η^δ with state space $[A, \mathfrak{A}]$, for each admissible strategy δ. For each admissible strategy δ let ξ^δ be the Markov process which is obtained from η^δ by addition of the actual time in the description of the state of the controlled system. So ξ^δ is a homogeneous MARKOV process with state space $[E, \mathfrak{B}]$, $E = R^+ \times A$, $\mathfrak{B} = \mathfrak{B}^+ \otimes \mathfrak{A}$. ξ^δ is to stop at time γ, where γ is defined as in A3; $\zeta \equiv \infty$, and T has the same meaning as in section 3. In the following our aim is to give conditions for the parameters of the controllable service model, which we have introduced above, such that this model realizes the assumptions of the lemma and of the theorems 1, 2 respectively.

Definition 2: An admissible strategy δ is called *right continuous* if

$$\lim_{\substack{t \searrow t_0 \\ a \searrow a_0}} \delta(t, a) = \delta(t_0, a_0)$$

for each $0 \leqq t_0 < T$, $a_0 \in A$, where for a sequence (a_m) of elements of A_g $a_m \searrow a_0 \in A_g$ means that $y_j(a_m) \searrow y_j(a_0)$ for each $j \in J_g$. At this $y_j(a)$ denotes the additional variable which belongs to O_j when the system is in state $a \in A_g$, $g \in G$, $j \in J_g$.

Now let the following assumption be satisfied for our model:

V: The set D of strategies consists of all right continuous admissible strategies.

By V it is clear that D is as in section 3 and the assumptions A1, A2, (I) are satisfied too. For each $(t, a) \in Z = [0, T) \times A$ we denote by $\xi^\delta_u(t, a)$, $0 \leqq u < T - t$, the value of the process ξ^δ at time u under the condition "$\xi^\delta(0) = (t, a)$, $\tau_1 > u$". (For the definition of τ_1 see p. 147.) $\xi^\delta_u(t, a)$ may be represented by $\xi^\delta_u(t, a) = (t + u, a^\delta_u(t, a))$. Furthermore let $\{t+u\} \times A^\delta_{u,s}(t, a)$ be the set of all states in which the process ξ^δ can be at time u under the condition "$\xi^\delta(0) = (t, a)$, $\tau_1 = u$, the object O_s produces the change of the coarse state at time u", $s \in J_g$, $a \in A_g$, $0 < u < T - t$. By the construction of ξ^δ the set $A^\delta_{u,s}(t, a)$ has finitely many states only. The set $A^\delta_{u,s}(t, a)$ consists of all such elements $a' \in A$ for which

a) $p(g(a), s, g(a')) > 0$

b) $j \in J_{g(a')} \backslash (J_{g(a)} \backslash \{s\}) \Rightarrow y_j(a') = 0$

c) $j \in J_{g(a)} \backslash \{s\} \Rightarrow j \in J_{g(a')}$, $y_j(a') = y_j(a^\delta_u(t, a))$.

Here $g(a)$ is defined by $a \in A_{g(a)}$. Because of the boundedness of each strategy $\delta \in D$ it is easy to show that $\boldsymbol{P}^\delta_x(\tau_i > 0) = 1$ for $i = 1, 2, \ldots$ and $\boldsymbol{P}^\delta_x(\lim_{i \to \infty} \tau_i = \infty) = 1$ for arbitrary $x \in E$ and each $\delta \in D$ if we assume that $T = \infty$ (see also [3], [4]). This means that in finite time only finitely many jumps of the coarse state of the process take place.

The probability $\boldsymbol{P}^\delta_{(t,a)}(\tau_1 > u) = \boldsymbol{P}^\delta_{(t,a)}(\xi(u) = (t+u, a^\delta_u(t, a))$ is given by

$$\boldsymbol{P}^\delta_{(t,a)}(\tau_1 > u) = 1 - \sum_{j \in J_{g(a)}} \frac{f_{i_j,j}(y_j(a_j))\, \delta_j(t, a)\, u}{1 - F_{i_j,j}(y_j(a_j))} + o(u)$$

and for a real measurable bounded function $f(t, a)$ we get

$$\begin{aligned} \boldsymbol{T}^\delta_u f(t, a) = \boldsymbol{E}^\delta_{(t,a)} f(\xi_u) = \boldsymbol{P}^\delta_{(t,a)}(\tau_1 > u)\, f(t+u, a^\delta_u(t, a)) \\ + \sum_{s \in J_{g(a)}} \sum_{a' \in A^\delta_{u,s}(t,a)} \bar{r}(a, s)\, p(g(a), s, g(a'))\, \delta_s(t, a) \cdot u \cdot f(t+u, a') + o(u) \end{aligned}$$

where

$$\bar{r}(a, s) := \frac{f_{i_s,s}(y_s(a))}{1 - F_{i_s,s}(y_s(a))} \quad \text{for} \quad a \in A, \quad s \in J_{g(a)} .$$

Therefore $A^\delta f$ is

$$\begin{aligned} A^\delta f(t, a) = \left(\frac{\partial f}{\partial t} + \sum_{j \in J_{g(a)}} \delta_j(t, a) \frac{\partial f}{\partial y_j} \right)(t, a) \\ - \sum_{j \in J_{g(a)}} \bar{r}(a, j)\, \delta_j(t, a)\, f(t, a) \qquad (5.2) \\ + \sum_{s \in J_{g(a)}} \sum_{a' \in A^\delta_{0,s}(t,a)} \bar{r}(a, s)\, p(g(a), s, g(a'))\, \delta_s(t, a)\, f(t, a') \end{aligned}$$

for all $f \in \boldsymbol{B}$ for which the right side of (5.2) exists. $\frac{\partial f}{\partial(\cdot)}$ means the right-side derivation of f. We get the elements of $A_{0,s}(t, a)$ from the elements of $A_{u,s}(t, a)$ by $u \searrow 0$. Furthermore we have

$$\lambda^{\delta}_{x,g} = \lambda^{\delta}_{(t,a),g} = \sum_{j \in g(a)} \delta_j(t, a)\, \bar{r}(a, j)\, p(g(a), j, g)\,.$$

Finally, in order to ensure that assumption A5 is valid we suppose that the functions $b_{i,j}(v)$, $j \in J$, $i \in C_j$, are right-continuous and that the functions $\bar{r}(a, s)$, $s \in J_{g(a)}$, are right-continuous in a: $\lim\limits_{a \searrow a_0} r(a, s) = r(a_0, s)$. This is sufficient because of the regularity of the temporal valuation of our model and assumption V respectively.

6. Example

The service system $GI/M/1/0$ with controllable speed for work of the service facility

Let a single server loss system with recurrent stream of arrivals be given.

We suppose that the distribution function $R(x)$ of the interarrival times is to have the density $r(x)$ and it is assumed that $R(x) < 1$ for all $x \in R^+$.

The quantity of work which is necessary for service of an incoming customer is exponential distributed with parameter u, $0 < u < \infty$. The speed for work of the server may be chosen from $V := [v_{\min}, v_{\max}]$, $0 \leqq v_{\min} < v_{\max} < \infty$, at each instant of time when the service facility is busy. If the server works with speed $v \in V$ for one time unit so the cost incurred in this time unit is $b(v)$, where b is a real-valued measurable bounded function defined on V. For each terminated service the system gets a finite reward w.

Finally, if an arriving customer finds the system busy so he leaves the system and simultaneously the system incurs the finite cost $k > 0$. Now we describe this model as controllable service scheme in the sense of section 5:

- $J = \{0, 1\}$,
- O_0: stream of arrivals, O_1: service facility,
- $C_0 = \{1, \bar{1}\}$, $D_0 = \emptyset$.

The object O_0 is in state 1 if the last incoming customer was accepted and is in state $\bar{1}$ if the last incoming customer was lost.

- $C_1 = \{1\}$, $D_1 = \{0\}$.

The service facility O_1 is in state 1 if it is busy and is in state 0 otherwise.

- $G = \{(1, 0), (1, 1), (\bar{1}, 0), (\bar{1}, 1)\} \simeq \{0, 1, \bar{0}, \bar{1}\}$

$g \in \{0, \bar{0}\}$ means that the system is empty and $g \in \{1, \bar{1}\}$ means that the system is busy.

- $V_{0g} = \{1\}$ for each $g \in G$.

This means that the realization of the interarrival times always takes place with speed 1.

$-V_{1g}=V$ for $g\in\{1,\bar{1}\}$

$-A=(\{0\}\times R^+)\cup(\{\bar{0}\}\times R^+)\cup(\{1\}\times R^+)\cup(\{\bar{1}\}\times R^+)$[2]

$-p(0,0,\bar{1})=p(\bar{0},0,1)=p(1,0,\bar{1})=p(\bar{1},0,1)=p(1,1,0)$
$\quad =p(\bar{1},1,\bar{0})=1$

$-k_{0,\bar{1}}=k_{\bar{0},1}=0;\quad k_{1,0}=k_{\bar{1},\bar{0}}=-g,\quad k_{1,\bar{1}}=k_{\bar{1},1}=k$

$-b_0(1)=b_{\bar{0}}(1)=0$

$b_1(1,v)=b_{\bar{1}}(1,v)=b(v)$

$-h(\cdot):\equiv 0$

$-f_{1,0}(x)=f_{\bar{1},0}(x)=r(x)$

$f_{1,1}(x)=\mu e^{-\mu x},\quad x\in R^+.$

The state $\bar{1}$ of the object O_0 is introduced only in order to guarantee that the new coarse state after a jump is different from the last coarse state. But it is clear that the optimal speed for work for the service facility may be chosen equal in the coarse states $1, \bar{1}$ of the system if an optimal strategy exists.

Analogous if δ is any strategy with $\delta(t,(1,y))=\delta(t,(\bar{1},y))$ for all $t, y\geqq 0$ then

$$\Phi(x,\delta)=\Phi(x',\delta)\quad\text{for all}\quad t, y\geqq 0$$

where $x=(t,(1,y))$, $x'=(t,(\bar{1},y))$.

Theorem: *Let D be a given set of right continuous admissible strategies. There exists an optimal strategy $\delta^*\in D$ if and only if there exists a bounded solution $f=$ $=(f_0, f_1)$ of*

$$\frac{\partial f_0}{\partial t}+\frac{\partial f_0}{\partial y}-\bar{r}(y)\,f_0(t,y)+\bar{r}(y)\,f_1(t,0)=0$$

$$\min_{\delta\in D}\left\{\frac{\partial f_1}{\partial t}+\frac{\partial f_1}{\partial y}-(\bar{r}(y)+\delta(t,y)\,\mu)\,f_1(t,y)\right.$$
$$+\bar{r}(y)\,f_1(t,0)+\delta(t,y)\,\mu f_0(t,y)$$
$$\left.+\bar{r}(y)\,k-\delta(t,y)\,\mu w+b(\delta(t,y))\right\}=0$$

$f_0(T,y)\equiv f_1(T,y)=0.$

If D contains at least all "constant" strategies then an assertion analogous to that of theorem 2 is valid.

Let now be $b(v)=v^2$ and $\Phi_0^\delta(t,y)$, $\Phi_1^\delta(t,y)$, $\Phi^\delta(t,y)$ defined by

$$\Phi((t,a),\delta)=\begin{cases}\Phi_0^\delta(t,y) & a\in A_0,\quad y(a)=y\\ \Phi_1^\delta(t,y) & a\in A_1,\quad y(a)=y\end{cases}$$

$\tilde{\Phi}^\delta(t,y)=\Phi_0^\delta\,(T-t,y)-\Phi_1^\delta\,(T-t,y)\quad 0\leqq t<T.$

---

[2] For simplification we neglect the quantity of work which the server has performed previously if the system is in coarse state $g\in\{1,\bar{1}\}$. Because of the known peculiarity of the exponential distribution this additional variable is not object.

Because of the lemma (see p. 149) we obtain the following differential equation for Φ^δ:

$$\frac{\partial\tilde{\Phi}^\delta}{\partial t}-\frac{\partial\tilde{\Phi}^\delta}{\partial y}=-\bar{r}(y)\,[\Phi^\delta(t,y)+k]+\mu\delta(t,y)\,w+\mu\delta(t,y)\,\Phi^\delta(t,y)+b(\delta(t,y))$$
$$\Phi^\delta(0,y)\equiv 0\,.$$

The solution of this equation is

$$\tilde{\Phi}^\delta(t,y)=\int_0^t[\mu\tilde{\delta}(\tau,t+y-\tau)\,w-b(\tilde{\delta}\,(\tau,t+y-\tau))-\bar{r}\,(t+y-\tau)\,k] \tag{6.1}$$
$$\times\exp\left\{-\int_\tau^t\mu\tilde{\delta}((u,t+y-u)+\bar{r}\,(t+y-u))\,\mathrm{d}u\right\}d\tau\,,$$

where $\tilde{\delta}(t,y):=\delta\,(T-t,y)$.

From (6.1) it follows that $\tilde{\Phi}^\delta(t,y)\leqq 0$ if $\bar{r}(y)\geqq\frac{\mu^2w^2}{4k}$ for all $y\geqq 0$.

On the other hand, because of (4.29) it follows that for an optimal strategy δ^* the relation

$$v_{\min}<\delta^*(t,y)<v_{\max}$$

is valid if $\Phi^{\delta^*}(t,y)\leqq 0$ and $v_{\max}$ is sufficient large.

Corollary: *Let $\tilde{\delta}(t,y)$ be a solution of*

$$\frac{\partial\tilde{\delta}}{\partial t}-\frac{\partial\tilde{\delta}}{\partial y}=\tilde{\delta}(t,y)\left[-\bar{r}(y)+\frac{\mu w}{2}-\mu\tilde{\delta}(t,y)\right]+\frac{\bar{r}(y)}{2}[w+k\mu]-\frac{\mu^2w}{2}$$
$$\tilde{\delta}(0,y)\equiv\frac{w}{2}\,,$$

where the function $\bar{r}(y)$ is assumed to be right continuous. If furthermore $\bar{r}(y)\geqq\frac{\mu^2w^2}{4k}$ for all $y\in R^+$ and

$$v_{\min}\leqq\sup_{0<t<T}\tilde{\delta}(t,y)<v_{\max}$$

then there is an optimal strategy δ^*, $\delta^*(t,y):=\tilde{\delta}\,(T-t,y)$.

References

[1] DYNKIN, E. B.: MARKOV Processes. Berlin, Göttingen, Heidelberg 1965.

[2] ФАЙНБЕРГ, M. A., и E. A. ФАЙНБЕРГ: Управление в системах массового обслуживания Зарубежная радиоэлектроника (1975) 3, 3–34.

[3] KÖNIG, D., K. MATTHES und K. NAWROTZKI: Verallgemeinerungen der ERLANGschen und ENGSETschen Formeln. Akademie-Verlag, Berlin 1967.

[4] KÖNIG, D., K. MATTHES und K. NAWROTZKI: Unempfindlichkeitseigenschaften von Bedienungsprozessen. Anhang in: GNEDENKO, B. W.; KOWALENKO, I. N.: Einführung in die Bedienungstheorie. Akademie-Verlag, Berlin 1974.

[5] РЫКОВ, B. B.: Управляемые системы массового обслуживания Теория вероятностей, Математическая статистика, Теоретическая кибернетика **12** (1975), 43–153.

[6] SIMONOVITS, A.: Direct comparison of different priority queueing discipline. Studia Scientarum Mathematicarum Hungarica **8** (1973), 225–243.

[7] SOBEL, M. J.: Optimal Operation of Queues. Lecture Notes in Economical and Mathematical Systems. Mathematical Methods in queueing theory **98** (1973), 231–261.
[8] STIDHAM, S., and N. U. PRABHU: Optimal control of queueing systems. Lecture Notes in Economical and Mathematical Systems. Mathematical Methods in queueing theory **98** (1973), 263–294.
[9] VERMES, D.: A necessary and sufficient condition of optimality for MARKOVian control problems. Acta Sci. Math. **34** (1973), 401–413.

Zusammenfassung

Es wird ein Steuerproblem für MARKOWsche Prozesse mit allgemeinem Zustands- und Entscheidungsraum und endlichem Steuerzeitraum untersucht. Der Wert einer gegebenen Strategie wird definiert als Erwartungswert des Gesamtertrages bis zu einem gegebenen Zeitpunkt. Die Erträge werden einerseits durch eine Ertragsrate und andererseits durch eine Bewertung der Sprünge des gesteuerten Prozesses festgelegt. Es wird eine notwendige und hinreichende Bedingung für die Existenz einer optimalen Strategie angegeben. Die Struktur der untersuchten Prozesse gestattet es uns, eine Klasse von Optimierungsproblemen der Bedienungstheorie als Spezialfall des allgemeinen Modells zu interpretieren.

Резюме

Исследуем проблему оптимального управления марковскими процессами на конечном интервале времени с общими пространствами состояний и управлений. Цена данной стратегии определенная как математическое ожидание суммарного дохода до данного момента времени. Доходы определенные с одной стороны плотностью дохода и с другой стороны оценкой скачков управляемого процесса. Даются необходимое и достаточное условие существования оптимальной стратегии. Структура исследованных процессов сделает возможным нам интерпретировать класс программированных проблем теории массового обслуживания как специальный случай общей модели.

Eingereicht bei der Redaktion: 4. 5. 1977; revidierte Fassung: 12. 6. 1978

R. Wets: **Grundlagen konvexer Optimierung.** Lecture Notes in Economics and Mathematical Systems 137, Springer-Verlag, Berlin-Heidelberg-New York 1976, 146 S., DM 18,–

Das vorliegende Buch stellt eine bemerkenswerte Zusammenfassung wesentlicher Resultate der konvexen Optimierung in übersichtlicher Form dar. Die einzelnen Kapitel lauten: I Konvexe Mengen und Konvexe Funktionen, II Stetigkeit, III Trennungssätze, IV Konjugierte Funktionen, V Darstellungssätze, VI Alternativsätze – Anwendungen, VII Differenzierbarkeit, VIII Projektionen und Schnitte konvexer Mengen, Funktionen und Subdifferentiale, IX Minimierung (konvexer) Funktionale, X Dualitätstheorie für konvexe Programmierungsprobleme, XI Algorithmen für konvexe Programmierungsaufgaben. In den Kapiteln I bis IV werden die bekannten Resultate über konvexe Punktmengen im R^n und über konvexe Funktionen, die den R^n in die Menge $R \cup \{\infty\} \cup \{-\infty\}$ abbilden, zusammengestellt. Die betrachteten Aussagen werden dabei weitgehend ohne Beweis angegeben. Im Rahmen des Kapitels III wird auf mehrere Trennungsbegriffe und wichtige Stützeigenschaften von konvexen Mengen, so u. a. auf den Satz von Minkowski, eingegangen. Kapitel V befaßt sich mit Extremalpunkten, Extremalstrahlen und exponierten Punkten. Hierbei spielen u. a. die Sätze von Carathéodory und Strazewicz eine wichtige Rolle. In engem Zusammenhang mit den Extremalpunkten steht die polyedrische Konvexität. Hier wird besonders auf die Sätze von Minkowski-Weil und von Farkas eingegangen. Kapitel VI enthält lineare Ungleichungssysteme und deren Anwendungen bei linearen Optimierungsproblemen sowie allgemeine konvexe Systeme. Dabei wird der Satz von Neumann zur Darstellung des Zusammenhanges der Lösbarkeit dualer linearer Optimierungsprobleme und deren Zulässigkeit herangezogen. Ferner werden die bekannten Zusammenhänge der Sätze von Helly und Carathéodory hergeleitet und der Minimax-Satz von Sion bewiesen. In Kapitel VII werden die bekannten Differenzierbarkeitsbegriffe dargestellt, d. h. Subdifferenzierbarkeit, Gâteaux-Differenzierbarkeit und Fréchet-Differenzierbarkeit. Neu in diesem Zusammenhang ist der Begriff der Hemidifferenzierbarkeit, der aber äquivalent ist zum Begriff der einseitigen Gâteaux-Differenzierbarkeit.

In Kapitel IX werden notwendige und hinreichende Bedingungen für die Minimierung ohne Nebenbedingungen hergeleitet. Die erhaltenen Resultate werden dann auf die Minimierung mit Nebenbedingungen, d. h. auf die Summe aus ursprünglicher Zielfunktion und der Indikatorfunktion der durch die Nebenbedingungen gegebenen Menge angewandt. Die Herleitung dieser Resultate erfolgt unter wesentlicher Verwendung der Hemidifferenzierbarkeit. Relativ breiter Raum wird der linearen Approximation der Nebenbedingungen eingeräumt, wobei eine Reihe von Regularitätsbedingungen betrachtet wird, mit deren Hilfe Kriterien für das Vorliegen eines Minimums hergeleitet werden können.

In knapper Form wird auf die Lagrange-Theorie und die sich daraus ergebenden Bedingungen für das Vorliegen eines Minimums eingegangen; die wichtigsten Resultate werden bewiesen. In diesem Zusammenhang wird auch die Kuhn-Tucker-Theorie für konvexe Optimierungsprobleme dargeboten. Kapitel X behandelt die Dualitätstheorie für konvexe Optimierungsprobleme. Bei der Formulierung des dualen Problems wird von geometrischen Betrachtungen ausgegangen. Es werden Bedingungen für schwache und starke Dualitätsaussagen hergeleitet. Der Leser wird in kurzen Zügen mit der Störungstheorie vertraut gemacht. Zusammenhänge zwischen Lösbarkeit, Dualisierbarkeit und Stabilität werden angegeben. Kurz wird die Beziehung zwischen Dualität und LAGRANGEfunktional dargestellt. Als Mangel empfindet man, daß die hier offensichtlich vorhandenen Verbindungen zu den im Kapitel IV eingeführten konjugierten Funktionen (d. h. die Zusammenhänge zum Dualitätssatz von FENCHEL) nicht aufgezeigt werden. Damit wird ein wichtiger Teil der Dualitätstheorie nicht behandelt. Kapitel XI beschäftigt sich mit Verfahren zur Lösung von konvexen Optimierungsproblemen, wobei die bekanntesten und wichtigsten Verfahren vorgestellt werden. Die unzureichenden Literaturhinweise empfindet man als Mangel. Dies ist besonders für Studierende von Bedeutung, wenn sie tiefer in das Gebiet der konvexen Optimierung eindringen wollen. Dessenungeachtet ist das Buch insgesamt eine interessante Bereicherung der einführenden Literatur in die konvexe Optimierung. Einige Unsicherheiten beim Gebrauch der deutschen Sprache beeinträchtigen die Nützlichkeit des Buches nicht.

K.-H. ELSTER

W. VOGEL: **Vektoroptimierung in Produkträumen.** Mathematical Systems in Economics 35, Verlag Anton Hain, Meisenheim am Glan 1977, 139 S., DM 28,–

Mit dem vorliegenden Bändchen wird erstmals der gegenwärtige Stand zu einigen Problemen der Theorie der (nichtlinearen) Vektoroptimierung in Form einer Monographie dargestellt. Der Verfasser stützt sich dabei im wesentlichen auf Resultate von BOL, GEOFFRION und KUHN/TUCKER, auf seine eigenen Ergebnisse sowie auf Untersuchungen seines Forschungskollektivs.

Das Buch gliedert sich in acht Kapitel; in den ersten vier Kapiteln werden nur halbgeordnete Mengen benötigt, während ab Kapitel 5 Mengen aus halbgeordneten (topologischen) Vektorräumen und den zugehörigen Dualräumen herangezogen werden.

Kapitel 1 (Geordnete Mengen und effiziente Punkte) enthält neben grundlegenden Begriffen allgemeine Aussagen über die Existenz effizienter Punkte sowie mehrere (geometrisch evidente) äquivalente Formulierungen für die Effizienz eines Punktes.

Kapitel 2 (Transformationssätze für geordnete Mengen) befaßt sich mit Aussagen über das Verhalten effizienter Punkte, wenn die Ausgangsmenge einer (monotonen) Transformation unterworfen wird. Damit erhält die bekannte Vorgehensweise der Einführung von Ersatzproblemen bei der Lösung von Vektoroptimierungsproblemen ihre Begründung; auf die Vielzahl bekannter Ersatzprobleme wird in den folgenden Kapiteln (teilweise) eingegangen.

Das Kapitel 3 (Effiziente Punkte in Produkträumen) spielt eine wesentliche Rolle für die folgenden Kapitel und enthält vorwiegend neue Ergebnisse des Autors. Insbesondere werden die Zusammenhänge zwischen effizienten Punkten von Mengen in Produkträumen und den effizienten Punkten gewisser Projektionen dieser Mengen auf Faktorräume geklärt.

Kapitel 4 (Anwendungen auf die Produktionstheorie) spezialisiert die bisherigen Ergebnisse auf ein Produktionsmodell, das auf GALE und OPITZ zurückgeht.

Kapitel 5 (Vektoroptimierung in einem linearen Raum) bringt die bekannten Beziehungen zwischen den BAYES-Lösungen (Lösungen des parametrisierten Ersatzproblems) und effizienten Punkten. Dabei wird gegenüber bisherigen Resultaten die Konvexitätsforderung abgeschwächt. Gleichzeitig werden die gewöhnliche lineare Optimierung und Quotientenoptimierung in die allgemeine Theorie eingeordnet.

Auch in Kapitel 6 (Konvexe Optimierung) und in Kapitel 7 (Optimierung in BANACH-Räumen) werden Bezüge zur klassischen Optimierung hergestellt. Außerdem werden durch Benutzung der Regularitätsbedingung von SLATER BAYES-Lösungen und Sattelpunkte (KUHN-TUCKER-Bedingungen) in Beziehung gebracht und über Tangentialkegel lokale Effizienzeigenschaften untersucht, wobei auch das Konzept von DUBOVICKIJ/MILJUTIN einbezogen wurde.

Kapitel 8 (Maximumsfunktionen und Statistik) widmet sich der Untersuchung von Stützfunktionen gewisser Mengen, deren Subdifferentiale und ihrer Anwendung in der Testtheorie.

Keine Berücksichtigung fanden die Ansätze zur Dualitätstheorie für Vektoroptimierungsprobleme, wie sie etwa von BRECKNER, BRECKNER/KOLUMBÁN, ISERMANN, NOGIN, OETTLI, RÖDDER und SCHÖNFELD vorgeschlagen wurden. Auch auf Bemerkungen zu Lösungsverfahren wurde verzichtet.

Durch diese Inhaltsangabe wird klar, daß dem mathematisch interessierten Leser, von dem zum Verständnis einige Kenntnisse der Funktionalanalysis und der Theorie der nichtlinearen Optimierung erwartet werden, mit diesem Buch einerseits eine gute Zusammenfassung und einheitliche Darstellung bekannter Sachverhalte und andererseits viele Anregungen für weitere Untersuchungen gegeben sind.

Etwas ungewohnt ist zunächst die Anordnung des Materials, da sich die Beweise jeweils konzentriert am Ende eines jeden Kapitels befinden. Einige Schreibfehler, die man jedoch stets leicht als solche erkennt, beeinträchtigen kaum die Lesbarkeit. Zwei zitierte Literaturstellen fehlen im Literaturverzeichnis. Offenbar handelt es sich bei SALZ 1975 (zitiert auf S. 58) um SALZ, W.: Die Untersuchung einer Vektoroptimierungsaufgabe, die zur Beschreibung kooperativer Differentialspiele dient. Dissertation, Bonn 1975, und bei LUENBERGER 1969 (zitiert auf S. 94) um LUENBERGER, D.: Optimization by Vector Space Methods. John Wiley & Sons, New York 1969.

R. NEHSE

K. BEER: **Lösung großer linearer Optimierungsaufgaben.** VEB Deutscher Verlag der Wissenschaften, Berlin 1977, 254 S., 40,– M

In der vorliegenden Monographie wird eine systematische Darstellung von Lösungsmethoden für große lineare, deterministische, endlichdimensionale Optimierungsaufgaben gegeben. Es werden spezielle Simplexmethoden, Iterationsverfahren und Dekompositionsmethoden betrachtet. Gegenüber früheren Publikationen zur gleichen Thematik wird hier ein breiteres Spektrum von Verfahren zur Bearbeitung großer linearer Optimierungsaufgaben vorgestellt. Ferner wird der Schwerpunkt der Untersuchungen auf die allgemeinen Prinzipien und die Theorie der Verfahren und nicht vordergründig auf die Darstellung von Einzelverfahren gelegt. Auf Grund dieser Vorgehensweise werden interessante und nützliche ordnende Ergebnisse erzielt. So ergibt sich, daß alle Verfahren, die bei Verwendung der revidierten Simplexmethode die spezielle Struktur der Restriktionsmatrix berücksichtigen, konkrete Realisierungen der Methode von KIM sind. Weiter zeigt sich, daß alle bekannten Dekompositionsverfahren spezielle Realisierungen der Methode der zulässigen Richtungen für entsprechend gewählte Ersatzaufgaben mit stückweise affin-linearer konkaver Zielfunktion über einer konvexen polyedralen Menge sind.

Das Buch wendet sich an Leser, die mit den wesentlichen Fakten der Theorie und der Lösungsverfahren der linearen Optimierung vertraut sind. Aus Gründen der Sprachverständigung sind jedoch die Begriffe, Aussagen und Verfahren, die häufig benutzt werden, im Kapitel 2 und in einem Teil des Anhangs zusammengestellt.

Im Kapitel 3 werden stückweise affin-lineare konkave Funktionen (SALK-Funktionen) untersucht. Es werden Eigenschaften solcher Funktionen bewiesen, Optimierungsaufgaben mit SALK-Funktionen als Zielfunktion und linearen Restriktionen formuliert, Optimalitätskriterien und Lösungsverfahren für solche Optimierungsaufgaben angegeben.

Das Kapitel 4 ist speziellen Simplexmethoden mit Berücksichtigung der Struktur der Restriktionsmatrix gewidmet. Insbesondere werden die Methode von KIM und die Methode der Pseudobasis dargestellt und auf verschiedene Strukturen der Restriktionsmatrix zugeschnitten. Dabei wird das folgende allgemeine Konstruktionsprinzip verwendet:

– Ausgehend von der speziellen Struktur der Aufgabe wird festgelegt, welche Teilinformation über die inverse Basismatrix statt dieser Matrix selbst benutzt wird.

– Es wird angegeben, wie mit Hilfe dieser Teilinformation und der Ausgangsdaten der Aufgabe die Schritte der Simplexmethode zu realisieren sind.

– Es werden Rechenformeln entwickelt, mit denen die zur nächsten Basislösung gehörende Teilinformation aus der aktuellen Teilinformation und den Ausgangsdaten ermittelt wird.

Ferner enthält dieses Kapitel noch Ausführungen über Iterationsverfahren der linearen Optimierung. Ausführlicher wird jedoch nur ein Gradientenverfahren betrachtet.

Den zentralen Teil der Monographie bilden die im Kapitel 5 enthaltenen Untersuchungen zur Berücksichtigung der Struktur der Aufgabe bei den Dekompositionsverfahren. Es werden zwei verschiedene Klassen von Dekompositionsverfahren betrachtet. Diese sind die Dekomposition mittels Zielfunktionssteuerung und die Dekomposition durch Ressourcenaufspaltung. Für alle Dekompositionsverfahren wird stets das folgende allgemeine Konstruktionsprinzip benutzt:

– Ausgehend von einer Aufspaltung der Ausgangsaufgabe in Teilaufgaben wird eine äquivalente Ersatzaufgabe mit stückweise affin-linearer konvexer Zielfunktion formuliert.

– Das Subdifferential dieser SALK-Funktion wird mit Hilfe der Teilaufgaben für den aktuellen Iterationspunkt beschrieben.

– Mittels dieses Subdifferentials wird eine zulässige Wachstumsrichtung der SALK-Funktion bestimmt.

Im Anhang werden unter anderem für einige praktische Probleme zugehörige mathematische Modelle entwickelt, die sich als strukturierte lineare Optimierungsaufgaben erweisen.

Die vorliegende Monographie liefert einen interessanten und wertvollen Beitrag zur Theorie und Praxis großer linearer Optimierungsaufgaben. Aus Platzgründen sind in ihm leider keine ausführlichen Untersuchungen über die Realisierung der Algorithmen auf Rechenautomaten enthalten. Gerade solche Aussagen und eventuell detaillierte Ablaufpläne könnten die praktische Nutzung der Ergebnisse wesentlich fördern und beschleunigen.

M. SCHOCH

J. DESLOUX and J. MARTI (eds.): **Numerical Analysis.** Proceedings of the Colloquium on Numerical Analysis 1976, Birkhäuser Verlag Basel und Stuttgart 1977, 248 S., sFr. 44,–

Der Sammelband enthält 12 Beiträge des Kolloquiums über Numerische Analysis vom 11. bis 13. 10. 1976 in Lausanne, wobei der erste Beitrag in französischer, die restlichen in englischer Sprache erschienen sind. In diesen Beiträgen kommen sowohl die theoretischen Aspekte als auch praktische Anwendungen der Numerischen Analysis zur Sprache.

P. M. LASCAUX (Villeneuve – St. Georges, Frankreich) wendet bekannte numerische Methoden, insbesondere Diskretisierungsverfahren, zur Lösung von instationären zweidimensionalen Ausflußproblemen mehrerer nicht mischbarer Materialien an und vergleicht sie untereinander. Er erläutert seine Ausführungen zum Schluß durch verschiedene Rechenbeispiele.

J. A. NITSCHE (Freiburg) erläutert am Modell der eingespannten Platte die Methode der finiten Elemente vom mathematischen Standpunkt aus. Zunächst wird in der Einführung der Übergang vom Differentialgleichungs-Randwertproblem zur Variationsgleichung, die den Ausgangspunkt für die Methode der finiten Elemente bildet, vollzogen. Der Verfasser erwähnt die klassische RITZ-Methode und weist auf die Schwierigkeiten derselben hin, die auftreten, wenn man für allgemeine Bereiche Approximationsfunktionen sucht, die die Randbedingungen erfüllen und die erforderlichen Differenzierbarkeitseigenschaften haben. Ferner werden die Finiten-Element-Funktionenräume definiert und die typischen Approximationseigenschaften ihrer Elemente angegeben. Im Falle der Anwendung spezieller finiter Elemente (Polynome fünften Grades) mit isoparametrischen Modifikationen (Bramble-Zlamal) werden bei Verwendung der Ritz-Methode qualitative Fehlerabschätzungen angegeben. Zu den gleichen Fehlerabschätzungen kann man gelangen, wenn man unter Verzicht auf die isoparametrischen Modifikationen sogenannte Randmodifikationen vornimmt, durch die die Approximationsfunktionen bei gekrümmten Elementen nahezu verschwindende Randwerte annehmen und zum ursprünglichen Funktional gewisse Randterme addiert werden. Weiterhin wird gezeigt, daß auch bei geringeren Regularitätsforderungen an den Raum der Approximationsfunktionen durch Wahl einer geeignet erweiterten Bilinearform Konvergenz erzielt und eine passende Fehlerabschätzung angegeben werden kann. Schließlich wird auf sogenannte hybride Methoden eingegangen, bei denen die beim Plattenproblem auftretende Differentialgleichung vierter Ordnung durch ein äquivalentes System von zwei Differentialgleichungen zweiter Ordnung ersetzt wird und dabei eine Reduktion der Regularitätsforderungen möglich wird.

H. Niessner (Baden) stellt einige Methoden vor zur Approximation von Funktionen mittels Summen von Exponentialgliedern (multiexponential fitting). Zwei dieser Methoden werden auf das praktische Problem der Analyse von Heizungskurven angewendet, bei dem nur eine positive Summe von Exponentialgliedern auftritt.

D. J. Randazzo (Basel) erörtert die wesentlichen Punkte eines neuen Verfahrens zur Berechnung von Funktionsnäherungen in Form einer Summe von Exponentialgliedern nach der Fehlerquadratmethode. Dabei wird das auftretende nichtlineare Normalgleichungssystem unmittelbar betrachtet und nicht linearisiert. Anhand einiger Rechenbeispiele erläutert der Verfasser seine Methode und vergleicht sie mit anderen üblichen Verfahren.

D. M. Davierwalla (Würenlingen, Schweiz) löst die zweidimensionale Neutronen-Diffusions-Gleichung mittels der Methode der finiten Elemente, wobei er lineare, quadratische und kubische Lagrangesche Elemente verwendet. Seine Ausführungen enthalten auch Bemerkungen zur rechentechnischen Realisierung des Verfahrens und erläuternde Beispiele.

J. M. Blanc (Schweiz) berechnet mit Hilfe eines Computers die elektrischen und magnetischen Felder, die bei der Herstellung von Aluminium in den elektrolytischen Zellen auftreten. Zur numerischen Lösung des elektrischen Problems wird die Methode der finiten Elemente angewendet, wobei Dreieckselemente erster Ordnung benutzt werden. Die Lösung des magnetischen Problems kann unmittelbar mittels Integrationen gewonnen werden.

R. H. Morf und E. P. Stoll (Rüschlikon, Schweiz) beschreiben stabile numerische Integrationsverfahren zur Lösung eines großen Systems von gewöhnlichen nichtlinearen Differentialgleichungen, die bei der Berechnung von Molekularbewegungen auftreten. Gelöst werden sowohl die reinen Newtonschen Bewegungsgleichungen als auch diejenigen Bewegungsgleichungen, bei denen Dämpfung und stochastische Kräfte auftreten.

L. A. Glenn betrachtet eine endliche Differenzenmethode zur Lösung von vieldimensionalen Problemen in der Kontinuumsmechanik. Das Verfahren wird angewendet, um die Entwicklung der Bruchzone in Glasblöcken vorherzusagen. Es wird gezeigt, daß die Resultate der Rechnung durch Experimente bestätigt werden.

S. Merazzi (Lausanne, Schweiz) und P. Stehlin (Stockholm, Schweden) stellen einige Resultate einer laufenden Arbeit über das viskoelastische Verhalten von zusammengesetzten Strukturen vor. Unter Beachtung des nichtlinearen Verhaltens der Strukturen und des viskoelastischen Materials treten verschiedene numerische Lösungsmethoden auf: Extrapolationsmethode, endliche Differenzenannäherung, Newton-Raphson-Iteration,

Approximation der Relaxationsfunktionen durch eine Summe von Exponentialgliedern. Rechentechnische Hinweise und einige Beispiele ergänzen die Ausführungen.

P. Bremi (Winterthur, Schweiz) beschäftigt sich mit der Kombination von analytischen und numerischen Berechnungsmethoden. Er zeigt anhand einiger interessanter Beispiele, wie man durch eine geeignete analytische Vorbehandlung die numerische Lösung von Problemen überhaupt erst möglich machen oder zumindest effektiver gestalten kann.

G. Wanner (Genf, Schweiz) betrachtet die Integration von steifen Differentialgleichungen und erörtert verschiedene numerische Lösungsverfahren, nämlich Kollokationsmethoden und Methoden von Rosenbrock. Er gibt verschiedene Stabilitätsdefinitionen an und beschäftigt sich ferner mit Fehlerabschätzungen und Schrittgrößensteuerung. Hinweise auf ein entsprechendes FORTRAN-Programm und zugehörige Literatur beschließen den Beitrag.

H. Froidevaux behandelt Lösungsmethoden für eine gewisse Klasse von nichtlinearen Randwertproblemen im R^n, die beispielsweise in der Mechanik der idealen kompressiblen Flüssigkeiten, in der Magnetostatik oder der Elektrochemie auftreten. Dabei gibt er die Variationsformulierung der allgemeinen Randwertprobleme an und beweist einige Theoreme, die sich mit der Existenz und Eindeutigkeit der Lösung befassen. Zur numerischen Lösung des Problems wird die Methode der finiten Elemente verwendet, die auf das Beispiel der kompressiblen Flüssigkeiten auch angewendet wird.

O. Fuchs

Bezugsmöglichkeiten der Zeitschrift „Mathematische Operationsforschung und Statistik – Series Optimization":
Bestellungen sind zu richten

– *in der DDR* an den Postzeitungsvertrieb, an eine Buchhandlung oder an den AKADEMIE-VERLAG, DDR – 108 Berlin, Leipziger Str. 3/4
– *im sozialistischen Ausland* an eine Buchhandlung für fremdsprachige Literatur oder an den zuständigen Postzeitungsvertrieb
– *in der BRD und Westberlin* an eine Buchhandlung oder an die Auslieferungsstelle KUNST UND WISSEN, Erich Bieber, 7 Stuttgart 1, Wilhelmstr. 4–6
– *in Österreich* an den Globus-Buchvertrieb 1201 Wien, Höchstädtplatz 3
– *in den übrigen westeuropäischen Ländern* an eine Buchhandlung oder an die Auslieferungsstelle KUNST UND WISSEN, Erich Bieber GmbH, CH – 8008 Zürich/Schweiz, Dufourstraße 51
– *im übrigen Ausland* an den Internationalen Buch- und Zeitschriftenhandel; den Buchexport, Volkseigener Außenhandelsbetrieb der Deutschen Demokratischen Republik, DDR – 701 Leipzig, Postfach 160, oder an den AKADEMIE-VERLAG DDR – 108 Berlin, Leipziger Straße 3/4

Zeitschrift „Mathematische Operationsforschung und Statistik – Series Optimization"

Herausgeber: Akademie-Verlag, DDR – 108 Berlin, Leipziger Straße 3–4; Fernruf 2236221 oder 2236229; Telex-Nr. 114420; Bank: Staatsbank der DDR, Berlin, Kto.-Nr.: 6836-26-27012

Chefredakteur: Prof. Dr. Karl-Heinz Elster
Verantwortlicher Redakteur: Dr. Rüdiger Reinhardt

Anschrift der Redaktion: Sektion Mathematik, Rechentechnik und ökonomische Kybernetik der TH Ilmenau, DDR – 63 Ilmenau, Postfach 327

Veröffentlicht unter der Lizenznummer 1586 des Presseamtes beim Vorsitzenden des Ministerrates der Deutschen Demokratischen Republik

Gesamtherstellung: VEB Druckerei „Gottfried Wilhelm Leibniz",
DDR – 445 Gräfenhainichen

Erscheinungsweise: Die Zeitschrift „Mathematische Operationsforschung und Statistik – Series Optimization" erscheint jährlich in einem Band mit 4 Heften. Bezugspreis je Band 120,– M zuzüglich Versandspesen (Preis für die DDR 80,– M); Preis je Heft 30,– M (Preis für die DDR 20,– M)

Bestellnummer dieses Heftes: 1093/10/1

Printed in the German Democratic Republic.

AN (EDV) 40234

Contents

Book Reviews

Bei Fragen zur Produktsicherheit wenden Sie sich bitte an:
If you have any questions regarding product safety,
please contact:

Walter de Gruyter GmbH
Genthiner Straße 13
10785 Berlin
productsafety@degruyterbrill.com